FORKS
IN THE
TRAIL

Jack, age 37, at the School of Forestry,
University of Massachusetts at Amherst.

FORKS
IN THE
TRAIL

A Conservationist's Trek to the Pinnacles
of Natural Resource Leadership

JACK WARD THOMAS

FOREWORD BY
Char Miller

A BOONE AND CROCKETT CLUB PUBLICATION

Missoula, Montana I 2015

Forks in the Trail

A Conservationist's Trek to the Pinnacles of
Natural Resource Leadership
By Jack Ward Thomas

Library of Congress Catalog Card Number: 2015940096
Paperback ISBN: 978-1-940860-14-5
Hardcover ISBN: 978-1-940860-20-6
e-ISBN: 978-1-940860-15-2
Published July 2015

Published in the United States of America by the
Boone and Crockett Club
250 Station Drive, Missoula, Montana 59801
Phone (406) 542-1888
Fax (406) 542-0784
Toll-Free (888) 840-4868 (book orders only)
www.boone-crockett.org

Printed in the U.S.A.

This book is dedicated to the memory of great mentors that showed me the way of a professional wildlife biologist over a sixty-year career.

Dr. James G. "Jim" Teer

My first boss and mentor in the old Texas Game and Fish Commission, taught me my trade as a wildlife biologist and generously included me as an author on what proved to be significant contributions to wildlife literature. Though I was too young and inexperienced for the job, he recommended me as his replacement when he resigned to return to the University of Wisconsin to complete his doctorate. I tried my best not to let him down. He was my vision of what a practicing wildlife biologist should be.

CONTENTS

Jack with his parents, Louise and Scranton, during a picnic at the Thomas farm, Handley, Texas, 1952.

FOREWORD

Jack Ward Thomas grew up listening to and learning how to tell stories. That's no surprise, for he came of age in the Cross-Timbers region of Texas during the Great Depression within a tight-knit family. Its leading voices included Thomas's grandfather (Big Dad) and father, mother, aunt, and uncle, who, by their words and actions, offered up an endless supply of lessons; some were apocryphal, others didactic and moral. As Thomas recounts in this fascinating memoir, their insights became his, just as their passions guided him to his own.

Consider one of the key questions that animates this autobiographical narrative—how did Thomas, a Texas A&M–trained wildlife biologist, become the thirteenth chief of the U.S. Forest Service, and thus one of the leading conservationists in post–World War II America? The answer, he suggests, lies in a single word: husbandry.

It was from Big Dad that he first heard the term, and when he asked what it meant, his grandfather responded by marrying genealogy and stewardship. "He said that he was Grandmother's husband and that meant that he was to care for her and protect her and their children—and their grandchildren." By extension, Thomas's father was doing the same for his spouse and their progeny, and in time the boy would become a man: "He told me that someday I would have a wife, and it would be my turn to be a husband." In time, too, he would embrace the equally binding—because analogous—obligation to sustain the physical environment. It was "the responsibility of all people to 'husband' what he called the 'good earth,'" and young Jack never forgot Big Dad's beguiling vision of the reciprocal relationship between humanity and nature.

Nothing is ever that simple, of course, as Thomas, a canny writer, knows full well. Still, he also recognizes just how influential the family circle can be, how it can shape, guide, and mold its children's beliefs and actions. Thomas did not become the head of the Forest Service because one day long ago he was schooled in what it meant to be a husband, but that moment nevertheless was formative.

Every bit as compelling are the episodes small and large that Thomas selects to define the arc of his life. The college acquaintance who offhandedly suggested that wildlife studies might be more appealing than veterinary science; the engaged faculty who took him under their wing; the old-timers who taught him about the inadequacies of a college education and the need to respect those on whose shoulders he stood. Most of all, there was Thomas's first boss at the Texas Parks and Wildlife Department, under whose tutelage he came to appreciate the underlying principles of his new profession: "First, our field was more about people management than it was about wildlife management. And, second, the most sensitive nerve in the human body runs between the heart and the pocketbook. The key to success was intertwining and balancing those two truths."

Finding that balance proved elusive, whether standing on the wind-swept Edwards Plateau of Central Texas or tramping through the wooded hills and hollers of West Virginia, scene of Thomas's first research job with the Forest Service. It was no easier to locate amid Oregon's rugged Blue Mountains or Washington's snow-capped Cascades, high country for which he would help enact far-reaching management policies. The same held true during the years he spent roaming the high-polished corridors of power in the District of Columbia.

Everywhere along this meandering path, Thomas was confronted with the limits of his knowledge, the steep learning curves he had to surmount, the humility he did not always remember to practice. Fortunately he was quick on his feet and possessed boundless energy. Yet Thomas also was willing to work hard, very hard, a commitment that paid off, whether riding the range, laying down transect lines, wrestling with a 200-pound white-tail doe in hip-deep snow, or schmoozing with mountaineers, arguing with county agents, testifying before congressional representatives, and advising a president or two.

He needed to be nimble, too, for Thomas has had a real knack for landing in the thick of controversy. How to best manage runaway deer populations, a dispute that continues to roil the body politic, was only the first of many charged moments he found himself in. When in the late 1960s turkey hunters sued the U.S. Department of Agriculture, and thus the Forest Service, for clear-cutting large swaths of the Monongahela National Forest, he was there. As urban sprawl bulldozed deep into once-rural landscapes, Thomas and his colleagues at the University of Massachusetts tried to green up the flattened terrain by calculating whether trees heightened property values (they did); that innovative investigation led to a cutting-edge collaboration with the National Wildlife Federation to encourage suburbanites to nurture backyard habitats. No sooner had he moved to the Pacific Northwest in the early 1970s than a two-decade-long brawl erupted over some of the region's endangered and threatened species, among them the spotted owl, bighorn sheep, and Chinook salmon. There the now-seasoned professional learned some dispiriting lessons about the politics of science and the science of politics, which he relearned after arriving in Washington in 1993; he has related this latter, troubling experience in engrossing detail in *The Journals of a Forest Service Chief* (2004).

Through it all, Jack Ward Thomas has tried to hold fast to the down-home advice he had absorbed as a child. "There are times to talk and times to listen," Big Dad once advised his grandson. "And remember, you're not learning a damn thing when you're talking." If ever pushed to the wall, however—which, as these pages relate, happened a lot over the course of Thomas's career—never let "your alligator mouth overload your hummingbird ass."

Those have been choice words to live by.

Char Miller
CLAREMONT, CALIFORNIA

Char Miller is the director of the environmental analysis program at Pomona College and the W.M. Keck Professor of Environmental Analysis. He is the author of *Gifford Pinchot and the Making of Modern Environmentalism*, *Public Lands, Public Debates: A Century of Controversy*, and the forthcoming *Our Land: America's National Forests, Wilderness, and Grasslands*.

PREFACE

The U.S. Forest Service (USFS) has been at the center of the American conservation movement for a bit over a century. During my tenure as chief of the USFS (1993–1996), the agency's headquarters in Washington, D.C., originally housed the U.S. Mint. The mid-nineteenth-century, four-story, red brick building with its clock tower has a character all its own that distinguished it from what became more standard super-sized government buildings. It was located kitty-cornered from the huge "government-ugly," larger-than-life building constructed during the Depression Era to house the U.S. Department of Agriculture (USDA), of which the USFS is, has been, and will likely remain something of a bastard child.

If you enter the front door and ask for the Chief's Conference Room, you will be directed up the winding staircase to the second floor. The conference room occupies what once was the office of the first chief of the USFS, Gifford Pinchot, now an icon of the American conservation movement.

In the corridor leading to the conference room are the portraits of the sixteen chiefs in the "Long Green Line" who headed the agency over its slightly more than a century existence. The faces universally exude authority and confidence. When I first saw those portraits, I wondered who these people really were—searching for a hint of a third dimension. What experiences had made them who and what they were? What forks in the trail had they encountered in their life journeys where choices had determined their, and America's, conservation's path?

Likely, the images portrayed by the portraits are incomplete—or misleading. I had known five of the men and women portrayed in those photographs. We started our careers as journeymen professionals in conservation—foresters, an engineer, a range scientist, a wildlife biologist, and a fisheries biologist. We worked in various jobs in the USFS and elsewhere—state agencies, other federal agencies, academia, and private enterprise—across the United States and overseas. Whatever fears and foibles we dealt with along the way disappeared, or were masked by continued success of the agency and the mythology that surrounds USFS chiefs. Those missing experiences hide that second dimension of just who we really were and the experiences that made us what we were to become.

After working ten years for the Texas Game and Fish Commission and its successor agency, the Texas Parks and Wildlife Department, followed by thirty years for the USFS, I was privileged to spend a decade as the Boone and Crockett Club's endowed professor in the College of Forestry and Natural Resources at the University of Montana in Missoula. The seniors and graduate students I taught in my course Case Histories in Conservation were intrigued by my "rest of the story" renditions of forty-plus years of professional real-life experiences.

Those stories helped them understand the "forks in the trail" that led to changes in the management of wildlife and wild lands and to shifts of perspective and approaches in the conservation arena. Some provided a glimpse into the complex decision-making processes involved. Others revealed the consequences—good and bad—of decisions made along the way. My personal "war stories" provide glimpses into the lives and personalities of the individuals involved—and the times.

> Stories have a way of connecting us. We may not have photographs of all of our ancestors, but we do have stories. Sometimes these stories can reveal things a two-dimensional

Jack's grandparents, Pearl and Arthur L. Thomas Sr., along with their bird dog, Handley, Texas, 1936.

image cannot. They can reveal a person's character, for example, because they look behind the image into the heart and soul.

—Joseph M. Marshall III, *The Day the World Ended at Little Bighorn—A Lakota History*

These stories span my fifty-five-year professional career to date. They hint at the second dimension of one of those chiefs of the USFS whose photo hangs in that hallway and reflect the vagaries in the lives of all those who stare down from those photographs. Foibles, fears, mistakes, adventures, misadventures, successes, failures, and comedies of errors and ego—all contribute to that additional dimension.

These stories demonstrate that those whose images adorn the walls of power common in government buildings were very human. During my life I had many adventures, good and bad; achieved beyond my wildest dreams; and, to my lasting chagrin, too often fell short. In the process I learned much; lost often but won some too; suffered the "slings and arrows of outrageous fortune"; helped foment some troublesome quandaries; and helped in the resolution of others.

I hope these stories provide a glimpse into the third dimension that goes with my photograph and hints at the third dimension that lies behind all the photographs that hang in that hallway. The forks in our trails, those vagaries of life, could have made our lives turn out far differently had the decisions made at those critical forks in the trail been otherwise.

I came from a long line of storytellers. I can vouch for the storytelling skills of my grandfather, Arthur "Arlie" L. Thomas, and my father, Scranton "Scrant" Boulware Thomas. I rely on their memories of my great-grandfather and great-great-grandfather in believing that the storyteller gene runs strong in my paternal bloodline.

My grandfather ("Big Dad") and father told vivid and spellbinding stories. I never saw a movie that could compare with sitting around a campfire after a day's hunt and listening to them spin yarns and filling in the gaps with imagination. My latent talents in storytelling were honed over the course of many decades spinning and hearing yarns around campfires in many situations in North America and in many foreign lands. From time to time I wrote down a story, usually first dredged up from memory and told around a campfire.

Good storytellers become actors portraying different characters and situations by use of accents, speech patterns, facial expressions, body language, changes in tone and volume, pregnant pauses, and other tricks of the storyteller's trade. Written stories employ analogous tricks with which I had but little experience. But, as I wrote, the more I remembered and the more I wrote.

As I write this in 2015, nearly a decade after rounding out an even fifty-year career as a professional conservationist, I am enjoying "retirement" in the Bitterroot Valley of western Montana. I busy myself—as opportunities arise and the mood strikes—with a bit of writing, consulting, speaking, and teaching. Some—most especially my wife Kathy—say such activities serve admirably to keep me off the streets and out of the pool halls!

During my professional career, I kept, off and on, detailed journals. The portions dealing with my service as USFS chief were the basis of a book: *Jack Ward Thomas: Journals of a Forest Service Chief* (University of Washington Press). The remainder of the unpublished journal entries are the source of many of the stories in this volume.

FORKS IN THE TRAIL

A budding hunter at age five, Dallas, Texas, 1939. This is one of the earliest photographs of Jack with a gun, shooting left-handed at this point.

CHAPTER 1

A FOUNDATION IS SET:
HANDLEY, TEXAS, 1939-1956

I was born in 1934 in Fort Worth, Texas, in the midst of the Great Depression, but I grew up in Handley, a one-stoplight town on Highway 80 between Fort Worth and Dallas. My times spent on my grandfather's farm outside Handley led to my lifelong fascination with hunting and fishing. That evolved into making conservation of natural resources a vocation and a career. It was my early experiences on that hardscrabble farm that guided me toward what I would become—it was the first fork in the trail of what would be a professional career.

PLAYING IN THE TENT

In late summer of 1939, three scrawny black-haired kids—my two cousins and I—were playing in the grass-bur-infested yard of Big Dad's rambling, weather-worn farmhouse some four miles outside of Handley. The sky was darkening off to the north as towering, roiling black clouds approached. My grandmother and Aunt Charlotte called us into the kitchen saying that we could "play in the tent."

The "tent" was made of wetted bed sheets draped over the kitchen table and weighted down—table and floor—with bricks. Aunt Charlotte handed us crayons and paper salvaged from brown paper grocery bags as she joined us in the tent. Very soon, as the howling wind and dust cloud arrived, the tent's walls dimmed to an eerie twilight and screen doors slammed—over and over—as the old farmhouse creaked and moaned. The tin roof made scary popping sounds as the temperature dropped when the dust cut off the summer sun and the metal contracted.

Nearly six hours passed before the wind ceased howling and Aunt Charlotte crawled out of the tent, held back the edge of the sheet, and beckoned us to follow. The sunlight was a strange yellow as it shone through the dust hanging heavy in the air. Within minutes, our eyes, noses, and throats were burning. Outside, heavier dust lay like a thin yellow-brown blanket. I remember the heavy dust on Big Dad's 1928 Chevrolet sedan—the old man's pride and joy. It was a sad and shabby sight.

I saw Big Dad walking down from the barn where he had sheltered the milk cows and then had stayed to calm and protect them during the dust storm. The cows were bawling in the barn. Puffs of dust arose from the old man's every step. A faded red bandanna that had covered his nose and mouth hung loose around his neck, and copious tears ran from his red eyes, cutting grooves in the dust on his face. He sat down on the front steps and stared toward the county road where a passing truck raised a huge cloud of dust as it passed.

I sat down beside my grandfather and leaned up against him—both scared and a bit bewildered. I began to cry and rub at my eyes. He put his arm around my shoulders, and then he too began to shake as he broke into tears. I had never seen Big Dad cry—and never would again. After all, Big Dads don't cry, not in Texas anyway.

Playing in the tent had been a strange but oddly pleasant interlude. Now, emerging from the tent, I knew something was terribly wrong. This was no game.

Later that evening, I had heard a preacher on my crystal radio set wonder if the "end times" were upon us. I asked Big Dad if this was indeed fulfillment of those ancient warnings. He scoffed at attempts to lay the consequences of human folly on God's actions and inactions. He said that was especially true when things happened that we humans didn't, or didn't want to, understand. He didn't believe that God had much of anything to do with it. Mankind had sown folly, and now we reaped the whirlwind, pure and simple. Maybe, he said, we were being taught an overdue lesson—if we were smart enough to learn. Maybe, if and when the rains came again, we would do a better job of husbandry.

I asked what "husbandry" meant. He was quiet for a time before he answered with an analogy. He was Grandmother's husband, and it was his duty to care for her and protect her and their children and grandchildren. He said that my dad was my momma's husband and it was his job to care for her—and for me. He told me that someday I would have a wife, and it would be my turn to be a husband. And, he said, it was the responsibility of all—husbands, wives, and children to "husband" what he called "the good earth." I remembered his words and came to consider it good wisdom that I pondered many times in the decades to come. Many years later, when I became a university professor late in my professional life, I recognized that as a "teachable moment."

This was only one of the many dust storms that roared down from the Great Plains during the Dust Bowl days of the 1930s. To make matters worse, environmental catastrophe was coupled with economic and social disaster to form the signature of the what was to become known as the Great Depression. Big Dad, though an avid reader of classical literature and history with an impressive vocabulary, talked—in words and ways I could usually understand and remember—about relationships between humans and the world around us. He seized upon the awful "teachable moment": the Dust Bowl, he said, was the inevitable result of inappropriate treatment of the land too long and too carelessly pursued.

As I grew older, I was often present when the USDA's county agent visited Big Dad's farm to talk over federal agricultural programs and conservation efforts. There were signs on fence posts around Big Dad's farm proclaiming it to be a "WILDLIFE REFUGE." Somehow, in spite of that, it was alright for Uncle Henry and me to hunt and trap on the farm. Big Dad always left some grain in the fields for the birds and left brush piles unburned to provide cover for bobwhite quail. He also let brush grow along some of the fence lines for that same purpose. When he harvested post oaks and blackjack oaks for fence posts and firewood, he did so in half-acre patches to create openings where cover and food plants for wildlife could flourish in the sun.

Uncle Henry Simmons, who lived on the farm with Aunt Charlotte, subscribed to several hunting and fishing magazines, which he read over and over again. Big Dad collected bulletins from the county agent, including publications on land management sensitive to the needs of wildlife. When I could read for myself, those publications became my treasured library and triggers for daydreams.

Bedraggled men often walked up to the farmhouse from the railroad tracks that lay a couple of miles to the south. The interacting forces of the Great Depression and the Dust Bowl had cost them their farms or their jobs—or both. Now they simply wandered in search of the future. They had little or no money, and many carried what they owned on their backs. Most seemed tired, hungry, dirty, and dispirited. Some knocked on the door and asked for a handout—but not many. Most offered work in exchange for a meal and a warm dry place to sleep and maybe a bath at the windmill. I don't remember any being turned away. I remember sitting cross-legged on the floor of the front porch and listening as Big Dad talked with such men. I couldn't understand most of what they talked about, but they didn't laugh much.

The years of my youth, one of the toughest eras in the history of the United States, encompassed the Great Depression, the Dust Bowl, and World War II. Those mean, terrible years seared the minds and souls of the survivors. Their altered attitudes and behaviors marked their children forever—including, I suspect, me and mine.

NO MAN WHO CLAIMS A BIRD

By the time I really knew my Big Dad, he was in his late seventies. He was a lean man, and his snow-white hair fringed a bald pate. He was long retired from being a conductor on the railroad that ran north out of Fort Worth. His retirement came early when he made a decision to take a passenger train across a trestle over the Red River—which marked the border between Texas and Oklahoma—in flood stage. The trestle collapsed. He and the engineer, fireman, and flagman were acclaimed heroes for rescuing many passengers at great risk to their own lives.

But, hero or not, he had made a disastrous decision and paid the consequences. Some said the experience and his "early retirement" left him a bit "touched in the head." Even if what some said about Big Dad "having a screw loose" was true, I never much noticed. He was so good to me that maybe I didn't want to know anything else.

To me, my Big Dad was a wonderful—though reserved— grandfather. He seemed to enjoy my company and treated me, at least to my mind, as a grown man. He read voraciously and read to me for hours at a time, even after I could read for myself, nearly every day that I was with him. His favorites were Gibbon's *Decline and Fall of the Roman Empire* and the King James version of the Holy Bible. While Big Dad wasn't big on formal religion, he loved the beauty, poetry, and wisdom—and pragmatic lessons—set forth in the Bible.

He told me that his father, back in Tennessee just after the Civil War, had taught him to read, using that very same Good Book and reading by the light of a coal oil lamp late into the night. He was to leave me that tattered old Bible. When I read from it now, nearly seventy years later, I imagine that I can hear his voice and feel his presence.

He died in his sleep when he was eighty-five and I was fourteen. I have thought of him—a little or a lot—nearly every day since. Though we were separated in age by more than seven decades, he was my first close friend and mentor at a time when I sorely needed both. He still serves as an anchor for me.

Big Dad couldn't, or wouldn't, drive a gasoline-powered vehicle. He did his farm work via a mule-drawn wagon or plow or mounted on a mule astride an old Civil War McClellan cavalry saddle. He owned a 1928 four-door Chevrolet, and my Aunt Charlotte was his chauffeur. He kept a few saddle horses for Aunt Charlotte and Uncle Henry (and sometimes me) to ride. And he sometimes boarded and trained horses for others.

Working with the county agent, he created a five-acre farm pond on a tributary to Village Creek, which emptied into the Trinity River en route to the Gulf of Mexico and which attracted mourning doves during the fall hunting season, ducks and geese in the late fall and winter; it also provided fair fishing for perch and catfish all year round. That didn't count the pond's service as a swimming hole for family and friends, which was most welcome in the dog days of summer before the advent of air conditioning.

But Big Dad's passion, so far as wildlife was concerned, was providing and managing habitat for bobwhite quail. He bred and trained bird dogs—really good bird dogs—for his and his clients' use in pursuit of those birds. No present-day wildlife biologist could beat him as a manager of quail habitat. And from time to time, he bred, raised, and trained short-haired pointers for sale.

I learned much from Big Dad about wildlife, hunting, fishing, wildlife habitat management, literature, and the art of spinning a yarn. He could make talking about hunting and fishing almost as much fun as actually doing it. Most of what I learned from Big Dad—which was much and varied—came from watching what he did, listening to what he said, and tagging along with him every chance I got to watch and learn. As far as Big Dad was concerned, I deserved the title "the great imitator." I walked like him, talked like him, and to some degree learned to think like him.

Big Dad was born in Tennessee just after the end of the great Civil War—which he called the "War between the States." He set out on the road to Texas when he was sixteen, carrying everything he owned in a blanket roll slung over his shoulder. Big for his age, he got a job helping to build the railroad south into Fort Worth. By the time he was twenty-one, he was a foreman over a fifty-man crew because, as he told it, he could "whip (he pronounced the word as *whop*) any man's ass" who chose to question his authority.

I later suspected there were likely other reasons—early maturity, leadership, intelligence, bearing, and forceful personality beyond his years—that had come into play. However, just maybe being able to "kick ass and take names" was then, and remains, a handy skill from time to time. He worked his way up to a permanent position with the railroad and ultimately became a conductor.

Big Dad was big on courtesy—Southern style. He removed his hat when he talked to ladies and tipped his hat to men. He afforded respect to all: man, woman, or child, rich or poor, black or white. He addressed most people by title. He called special friends and kinsmen by their given names or, rarely, by a nickname. He called me "Grandson"—always.

I often tagged along when he trained bird dogs, hunted, or took

Arthur L. Thomas Sr., or "Big Dad," in his customary attire.

"dudes" hunting for a modest fee. When he had me in tow, Big Dad usually went into his teaching mode, and I usually took the lessons to heart. Looking back, nearly seven decades later, I can see that it sometimes took some time, even years, for me to fully grasp the lessons he was patiently imparting.

The year I was twelve, just before Christmas, Big Dad's much younger brother, Knox (who was in the dental supply business in Dallas), asked a favor of the old man. Would Big Dad take a special client of his quail hunting? That late in the quail season, Big Dad and his clients had already taken his self-imposed limit of quail for the season, and his brother knew it. But as it was Brother Knox doing the asking, the old man reluctantly agreed.

The dude—Big Dad referred to the "gentlemen hunters" who were his clients as "dudes"—arrived at the appointed hour in a big black shiny Packard, driven by a chauffer. Dude was dressed in a outfit right out of Montgomery Ward's ("Monkey Ward's" to us) catalogue: lace-up high boots, boot pants, a red-and-black-plaid wool jacket, and a new narrow-brimmed Stetson. He sported a neatly trimmed mustache and spoke with a distinct New York accent—loudly, fast, and a lot.

The image worsened when he pulled a brand new automatic shotgun out of a leather case. Big Dad firmly believed that gentlemen hunters used only shotguns with two barrels—side-by-side was preferred to over-and-under—in pursuit of bobwhites. Tradition!

We set out from the house with four dogs. Bo and Jangles walked at Big Dad's heels; I had Alphonse and Gaston on a leash and trailed behind. As we walked through the stubble fields bordered by post oak and blackjack oak forests, Big Dad began to lay out the standard operating procedures. Dude held up his hand and informed Big Dad that he neither required nor desired any instructions relative to hunting protocol. The old man didn't much cotton to either his words or tone but shrugged off the arrogant discourtesy.

When working dogs, Big Dad carried a double-barreled 20-gauge shotgun broken open and hanging over his arm. Even though he was not hunting, there were apt to be copperheads, rattlesnakes, and varmints of various types that might require killing.

Within a quarter hour, we flushed a covey of bobwhites. Big Dad kept the dogs at heel as this was one of his "house coveys," which he could almost always easily find and use to provide training opportunities for the dogs. Besides, he loved to hear the birds calling to one another close by the house, especially early in the morning and at dusk. His voice loaded with irritation, Dude demanded to know why he wasn't allowed to shoot.

Big Dad explained his reasons—rather patiently, I thought. Dude snorted in disagreement. He was not endearing himself to the old man, which I thought was apt to be a big mistake.

When we reached what Big Dad deemed more appropriate hunting territory, he gave a toot on his whistle and signaled the dogs to take up hunting. In less than ten minutes, Jangles locked down on point—which Bo honored. Big Dad motioned to Dude to walk forward between the dogs. When the bobwhites exploded into flight, Dude's first shot folded a bird flying straight away. The second shot was a clean miss. The third shot winged a bird angling off to the left, which spun to ground.

Dude must have assumed that Big Dad had also fired—which he hadn't. He hollered emphatically, "My birds! My birds!"

Big Dad called Jangles and Bo to heel, turned, and, without a word, started back toward the house. He jerked his head, signaling me to follow. Dude demanded, "Where are you going, Old Man?" Then, "What the hell is the matter with you?"

Big Dad turned and looked Dude square in the face. "Sir, I will hunt with no man who deems it necessary to claim a bird." Dude was left standing there slack-jawed to find his birds—all two of them.

I had turned the dogs into their pens at the house when Dude stormed up. He walked up to Big Dad and asked in a snotty tone, "Well, what do I owe you?" He was pushing the old man—not a really good idea, I thought!

"Sir, I neither require nor desire anything from you beyond your immediate departure from my property." And that was that. Dude, thank God, discerned that his best next move was to keep his yap shut and head for his car.

After the big black Packard disappeared, raising a cloud of dust all the way to the county road, Big Dad looked right at me and said, "Son, there is much more to hunting than killing—especially when hunting quail. There is a beauty to hunting quail over fine dogs that is sweetened by appropriate gentlemanly behavior—and made bitter by boorish behavior."

He walked to his wicker rocking chair on the front porch and eased down in respect for his lumbago. He packed his corncob pipe with Granger pipe tobacco out of a dark blue canister with the picture of a short-haired pointer frozen on point, lit it with a kitchen match, blew out a cloud of smoke, put his feet up on the porch railing, and leaned back.

He looked at me quizzically from under his greasy Stetson with a white bushy eyebrow raised as a question. I nodded—lesson learned.

GETTING SHOT

Big Dad's farm was located in the Cross-Timbers region of Texas between the Piney Woods to the east and the Prairies to the west. Post oaks and blackjack oaks struggled against Big Dad's sustained efforts to recapture fields so hard won by cutting the trees for fence posts, firewood, and building logs and then grubbing out the stumps—with the aid of a little mule power—to create crop and pasture lands.

Two huge pecan trees thrived in the bottomlands along an intermittent stream—Jack's Creek, where I caught crawdads and which fed into Village Creek and then on to the Trinity River and the Gulf of Mexico. Several intermittent creeks had been dammed, courtesy of a Depression-era government conservation program, to create three two- to five-acre impoundments. Those ponds—"tanks" in Texican—provided water for cattle, swimming, fishing, and hunting for doves and waterfowl.

I came to understand that Big Dad was decades ahead of his time as a manager of habitat for bobwhite quail. He applied his ideas, gained through observation and trial and error, to the interactions between habitat manipulations and bobwhite quail abundance. In the late 1940s and early 1950s, making money from managing land for the benefit of bobwhite quail and catering to hunters was a relatively new thing.

He told me that making money while having a good time was something few people ever managed to achieve. I never forgot that and vowed to search out a profession that would do that for me, though I knew nothing about "wildlife management" at the time.

During quail-hunting seasons, when I was handy—that is, not in school or able to escape therefrom—I served as Big Dad's "dog boy." My job was to follow the hunting party with two bird dogs on leash while carrying extra shotgun shells, lunches, and water for the hunters and dogs in bulging bags carried over each shoulder. When the dogs that were hunting became tired or overheated, Big Dad called them to heel, leashed the pooped pups, and signaled me to release the fresh pair of dogs. Then, with a toot on Big Dad's whistle, the hunt resumed.

One of Big Dad's safety rules, carefully explained to all hunters, was that under no circumstances was a hunter to shoot at quail that flushed back over the dogs and handler, who usually trailed along at thirty yards or so. From my assigned position I could watch and learn about hunting, handling dogs, dealing with hunters, and shooting.

During a severe (by Texas standards) cold snap in late December of 1947, Big Dad arranged with Dad for me to be available for a couple of days of quail hunting. I was 13 years old. His clients—longtime customers of Big Dad's brother Knox—were two Yankee dentists attending a convention in Dallas.

My grandmother had insisted that I dress to withstand the bitter cold. I donned long johns, denim jeans, T-shirt, wool work shirt, sweater, lined leather gloves, scarf, and a way-too-big, greasy, work-worn Stetson over an old navy-issue wool watch cap pulled down over my ears. I waddled when I walked—a toasty warm waddle!

The early-morning bone-chilling north wind blew in gusts, which encouraged the bobwhites to sit tight in their coveys and made things difficult for the dogs, as their scent was subdued, dissipated, and scarce. We began the day by hunting for over an hour with the dogs ranging wide and encountered only one covey. Ever-colder and increasingly discouraged, the dudes began to grumble.

As the sun rose to its zenith, the overcast dissipated, the day warmed a little, winds died away, and quail began to move about to feed. The dogs located and pointed several coveys in quick succession. Each of the hunters killed birds on the covey rises. Then the dogs pointed several singles, of which the dudes killed a half dozen or so. Their moods brightened along with the sun and rising temperatures.

As morning wore on, my layers of clothing became too much. I removed my heavy wool jacket and tied it to a shoulder bag. My wool shirt was unbuttoned, my gloves were stuffed in a back pocket, and the stocking cap was rolled up and stuffed in another pocket.

When we reached the farm's north boundary, we looped around to hunt back toward the farmhouse. Just a bit after midday, Big Dad called a short break and built a warming fire. I laid out sandwiches, a thermos of coffee for the hunters, and water for the dogs. As we were eating, the north wind picked up; the temperature dropped; and sleet spat off and on from the darkening sky. Piece by piece, I put back on the clothing I had removed earlier. My last move was to pull the wool cap over my ears and forehead and tug the old Stetson down over the cap.

At Big Dad's signal, I released two fresh dogs—Blitz and Baron—and leashed Alphonse and Gaston. When the old man tooted his whistle, we fanned out into hunting formation and worked our way back toward the farmhouse. The dogs were ranging wide at a lope.

We had not gone a quarter mile when Blitz went down on point and Baron backed. Big Dad eased in, whispering, "Hold, hold, hold," while signaling the hunters to fan out to either side. I was properly positioned thirty yards behind when a half-dozen quail flushed—almost straight up—and then turned back with the wind over the dogs and hunters.

I thought, "Damn! The hunters won't be able to shoot." What?! Both hunters pulled down on the birds that were flying directly toward me. The muzzles of the shotguns loomed huge and black. Big Dad's deep voice boomed, "No shot! NO SHOT! GAWD DAMN IT!!"

I ducked my head until my chin was on my chest and leaned forward just as the birdshot struck me full in the chest and head. Had I been shot? Yes! At least twice! My knees buckled. I sat down cross-legged, still clutching the dog's leashes.

Suddenly Big Dad was there. He knelt down beside me and pried the dog's leashes out of my hands. The look on his face did nothing for my peace of mind. He grabbed my upper arms and looked me straight in the eyes. "Are you hurt?"

I was plenty scared but didn't know if I was seriously hurt. "They shot me!" I was confused and scared. "Why did they shoot me?"

Big Dad pulled off the Stetson and the stocking cap and opened my pea coat. He saw blood staining my chin, lips, and teeth but none elsewhere. Big Dad assured me, "You'll be all right."

I wasn't sure I believed him—and I wasn't sure he believed himself.

He laid me back on the ground and unbuttoned the heavy sweater. Some birdshot fell from somewhere. My T-shirt had no blood on it. But where was the blood in my mouth coming from?

I sat up and spat blood and several pieces of shot from between my teeth and gums and lip. Some days later, two more pellets worked their way out of entry wounds on my chin. Lead marks on the crown and brim of the Stetson—which had covered my eyes and upper face when I ducked my head—indicated that my hat had turned away most of the shot. As near as Big Dad could reckon, I had been struck by two charges of shot—one centered on my chest while the other struck my upper chest and head.

The two hunters were horrified and tried to make amends. Big Dad was furious, likely even more scared than angry. When we reached the farmhouse, he took their payments for the hunt and curtly sent them on their way. He turned me over to my Aunt Charlotte to make sure I had a hot bath and then left to take care of the dogs. She probed the entry wounds but encountered no embedded pellets.

That night another "blue norther" arrived just at dusk as we sat around the woodstove toasting our shins. But unlike other such nights, very little was said. As I watched the smoke from Big Dad's corncob pipe ride the heat from the woodstove toward the ceiling, I relived the day's events. Surely, what had happened—which was plenty scary enough—was nothing compared with what might have been. My God, what if I had been ten or fifteen yards closer to the guns? What if I had not ducked my head at just the right time? What if the day had been mild and I was wearing only a ball cap and one layer of clothing? What if we had not stopped for lunch and I had not put my sweater, jacket, cap, hat, and gloves back on? What if?

"What if" could have entailed the loss of an eye, or eyes, or even my life—instead of a yarn to be told around future campfires. When I told the story over the years to come, at the conclusion I was nearly always asked, "What if?" Likely, many a fork in the trail would engender that question: "What if?"

BEST-LAID PLANS

In 1945–1946, a polio epidemic swept across North America. Just how the disease was spread was uncertain. Theories ranged from transmission by mosquitoes to person-to-person contact. Reactions by authorities ranged from broad-scale application of DDT to kill mosquitoes and other biting insects to isolation of victims exhibiting high fever and some degree of impairment of movement on the theory that they were contagious.

In the late summer of 1946, I came down with a high fever, violent headache, arched back from spasms, difficulty in breathing, and "funny feelings" in my legs. The local doctor was summoned, and he performed a cursory exam, bundled me into the back seat of his big black Buick, and delivered me to the isolation ward for polio patients at City-County Hospital in Fort Worth.

In spite of the fact that I wanted to believe that it was all a horrible nightmare, it was no dream. The influx of patients, mostly youngsters, exceeded the bed space, specialized equipment, and personnel available. I was placed on one of dozens of folding canvas cots that lined the hallways in the polio ward. The caregivers could not really do much for us beyond trying to keep us clean, fed (those who could swallow), and reasonably comfortable—and dealing with fear so raw that it seemed to me to give off a strange odor. Some had trouble breathing due to paralysis of the diaphragm and survived in what was called, euphemistically, an iron lung. No visitors were allowed.

As soon as the initial raging fever passed, it was assumed that patients were no longer contagious and could be transferred to other facilities or sent home. After several weeks, I was fitted with braces and sent home to recover as best I could with whatever help my folks and family doctor could provide. I was one of the lucky ones who suffered no long-term paralysis.

Most of us not permanently paralyzed were left to our own devices to recover as best we could. Many victims experienced what would now likely be called post-traumatic stress disorder, which, for me, was manifested in so-called "night sweats and terrors" that recurred for years. Though I recovered, I was not immune from the recurrent terrors that have returned from time to time over the years since.

For young men growing up in Texas in those days, especially those a bit bigger and a tad faster than average, no general societal expectation was more common than that of playing football, Texas style. High school football was then, and remains today, the unofficial religion of the sovereign state of Texas—the "Church of Friday Night Lights."

I worked hard and persistently to overcome the aftereffects of polio. I walked or ran every single day—farther and farther and faster and faster. I carried a pack loaded with bricks back and forth to school. I dreamed of playing football. Whether or not any of that helped, I recovered from the aftereffects of polio and went on to be become something of an all-around high school athlete. I set a record for varsity letters earned at my high school, which stood until the school was closed several decades later: four in football, four in basketball, and two in track. I also played baseball and softball and boxed in other venues, though they were not recognized varsity sports at the time.

Football was my game. As a freshman, I stood six feet two inches tall and weighed 185 pounds. By the time I was a senior, I weighed nearly 200 pounds, which was considered big in those days. I was also relatively agile, sort of fast, and had good hands and a sense for the game.

When I was a sophomore, I suffered a knee injury that was initially diagnosed as a simple torn cartilage, which was surgically removed (the standard procedure of the time). Then I was pushed back into action, likely much too quickly. Later it became clear that, in addition, I had suffered a torn anterior cruciate ligament (the dreaded ACL tear, in football talk).

I played three seasons after that injury—and never missed a game even with a knee that was held together with a special brace and what seemed like several thousand yards of adhesive tape over time. The knee injury cost me a little speed and a bit of agility, which can make all the difference at the top levels of competition. Still, I was a good enough to attract recruiters from Texas Christian University, Baylor, the University of Texas, Texas Tech, and Texas A&M. The recruiters reviewed game films, and several invited me to visit their campuses. But in the end I was simply not good enough for them to gamble a full-ride scholarship on a prospect with a bum knee. That was the end of my Plan A to pay for college—a big-time fork in the trail.

In retrospect, that injury was perhaps a good thing. I was good enough, big enough, fast enough, and gutsy enough to play big-time college football, at least in my opinion. But realistically I don't believe that I was bright enough or dedicated enough to be both an outstanding science major and a football player. Though I had taken all of the "hard courses" available at my small high school and had done well, I was not in competition for class valedictorian. All my teachers, save one, clearly considered me more a talented jock and a light-hearted jokester than a potential scholar.

Now what? I certainly did not qualify for an academic scholarship, and my parents were not low enough on the economic ladder for me to qualify for student aid. There were no government-guaranteed student loan programs in those days, at least none that I knew of. My folks could provide only minimal assistance. The Korean War was winding down. I was in the U.S. Navy Reserve and a call to active duty was certain unless I was granted a delay to attend college.

My English teacher, Miss Arabella Odell, was also the guidance counselor for seniors. She was the primmest and most proper of all my teachers—ever. We jocks, behind her back, called her "Momma Odell." She was, bar none, the best teacher I have ever known, and I held her in the highest esteem—and still do, three college degrees later. But she scared the hell out of me, all ninety-five pounds or so of her.

She asked, "Why not consider Texas A&M? Your grades are good enough for admittance. Uniforms are provided. Room and board and laundry are austere and plain but amazingly cheap. If you qualify for continuation in ROTC after your sophomore year, the military will subsidize your last couple of years and pay you a monthly stipend. With what you can earn doing construction work in the summers, you can make it if you suck it up and try your best!"

Why not? So I enrolled at what was then an all-male military academy and, to my surprise, thrived on the discipline, became a leader in the Corps of Cadets, and graduated on schedule in 1957—with honors, no less—with a degree in wildlife management and a regular commission as a second lieutenant in the U.S. Air Force. And I didn't owe anybody a dime. That was most fortunate, as I literally didn't have a dime.

I planned to parlay my military officer's commission into a career. While I waited for assignment to active duty, I found a job with the Texas Game Department (TGD) as a wildlife biologist headquartered at Sonora, where the Texas Hill Country meets the Chihuahuan Desert. Six months later, when I was ordered to active duty in the U.S. Air Force, a panel of flight surgeons evaluated X-rays of my knee, jerked my leg around, and gave a thumbs-down for flight training; I was relegated to the stand-by reserves. Fortunately, my temporary position with TGD was converted to a regular full-time appointment, and there was no turning back.

So that knee injury changed the course of my life for the second time. And it wasn't over. During the next half century, I would undergo three more surgeries on that knee, ending with a total knee replacement in 2007. When I look back over the changes in my life emanating from a simple football injury, I am struck by how much life can be changed by a single unanticipated event. Were those changes for the better or for the worse? Or was it just the way things were? That set me to wondering about forks in the trail and taught me to know a fork when I saw one emerging out of the fog of the future and how, just maybe, to turn it to best advantage.

FOLLOWING UNCLE HENRY

During World War II, Dad's sister, my Aunt Charlotte, and her husband, Henry Simmons, lived on Big Dad's farm. Aunt Charlotte had

Jack in his football uniform, Handley High School, 1949.

a beautiful warm spirit and sparkling eyes that directed attention from a disfiguring birthmark on her cheek and a forearm that was crooked from a poorly healed break that happened when she was eight years old. Uncle Henry stood a little over six feet and was, as Big Dad described him, a "tallish, scraggly, bald, and thinly" man. Uncle Henry was a great storyteller, especially when it came to hunting or fishing tales. He could, as Big Dad said, "tell it better than it really was—and that's no easy chore!"

Uncle Henry was, when the mood struck him—which was not as often as Aunt Charlotte would have liked—a master plumber. However, he was always in the mood for, as he put it, a "tad" of fishing and hunting.

Most of the young men who likely would have been hunting and fishing in those years were off fighting the war or working in the ship and aircraft factories. Those who were left were working long hours, six days or more a week, leaving little time for such activities as hunting and fishing. And to make hunting even more unlikely, sporting ammunition was rationed by government edict and was in short supply.

So when those facts were added up, there was much less hunting and fishing going on, so almost no attention was being paid to the enforcement of laws related to hunting and fishing. Uncle Henry had never been inclined to pay too much attention to legal constraints on his fishing and hunting activities in any case.

Some foodstuffs were rationed, especially meats of all kinds. My Uncle Arthur ("Ikie"), Big Dad's oldest son, ran a dairy farm on the north end of Big Dad's place. That gave our extended family a leg up when it came it came to obtaining rationed goods. From time to time, there were steers, bull calves, and "over age in grade" dry cows to butcher. Beef could be bartered for other items in either short supply or rationed. As a result, Uncle Henry and I were not troubled by any lack of ammunition.

Uncle Henry was, as Aunt Charlotte referred to him, a "hunting-fishing fool." Few days passed that he didn't do a little of what today's anthropologists would call subsistence hunting and fishing. Most days, well before daylight, Uncle Henry was up and dressed and "jazzed" on several cups of stout Folgers coffee, another severely rationed item obtained from astute trades. That routine allowed him to get in several hours of hunting or fishing before going to work.

His weapon of choice for hunting was a Remington single-shot bolt-action .22 caliber rifle. Uncle Henry was the best game shot I've ever known—and that's saying something. He missed but seldom and could, with some degree of regularity, shoot bobwhite quail on the wing with his rifle.

He trapped far more wildlife than he shot—bobwhite quail, mourning doves, fox and gray squirrels, opossums, feral pigeons, raccoons, cottontail rabbits, armadillos, and swamp rabbits. The beef, pork, and chickens raised on the farm were too valuable as barter if there was wildlife to take their place on the family's table. I was told that the game "tasted like chicken" no matter the species. All, including me, pretended to believe it.

Each species required a different trapping technique. Uncle Henry constructed most of his traps except for the leg-hold traps, which were "store-boughten." Pelts of opossums, raccoons, and rabbits taken during winter when their pelts were in prime condition were stretched and sold or traded to the fur trader, who, by chance, owned the local hardware store. He took in the furs in exchange for .22 rim-fire cartridges and shotgun shells. How he "cooked the books" to satisfy the inspectors we didn't know, care, or ask about.

Uncle Henry hunted waterfowl using a single-barrel 12-gauge shotgun with a Damascus twist-steel barrel designed for use with black powder cartridges. A few years later, I would "unravel" the barrel by firing off one too many more powerful "modern" shotgun shells loaded with smokeless powder.

In the chicken yard he kept several domesticated mallard ducks, originally brought down with broken wings. He had skillfully amputated the damaged wings. He called the ducks, poetically enough, "Judas birds." During waterfowl migrations, the Judas birds, anchored in place in one or another of the farm's ponds by a weight tied to a foot, lured in wild ducks to Uncle Henry's gun and sometimes mine. The Judas birds were decidedly more effective than the wooden decoys Uncle Henry skillfully carved and painted.

Thirteen-year-old Jack at fishing camp with the first gun he ever owned, a lever-action Model 39a Marlin .22 rifle. All the boys had brought their rifles in case there was time to do some squirrel hunting.

Uncle Henry unilaterally closed his hunting season from April through September to allow for wildlife to reproduce. During those months, fishing served to satisfy his predatory instincts and supplement the family's larder.

There weren't many fishing techniques, legal and illegal, that Uncle Henry had not mastered. I suspect now that he may have invented some of those techniques. Once I started second grade, I tagged along with Uncle Henry as often as possible as he went about his hunting, fishing, and gathering. He taught me much from his bag of proven hunting and fishing tricks. More important, he taught me to shoot—and shoot quite well—with both a shotgun and a rifle.

On my tenth birthday, Dad and Big Dad made arrangements with the friendly local hardware man to obtain a brand-new, straight-out-of-the-box, lever-action .22 caliber rifle—a Marlin Model 39A. Pulling that off in 1944, when there had been no firearms manufactured for civilian use since 1942, was likely not an easy matter. I have no idea how much fresh beef, pork, chickens, or dozens of eggs were required to make that deal. With that rifle, I joined a fraternity of fathers, grandfathers, uncles, and cousins who hunted. I was deemed ready, after years of apprenticeship, to be part of the brotherhood. On that day, I was a man—or at least well on the way. And I *felt* it; I *knew* it.

I recall the day I knelt beside the first animal, a swamp rabbit, that I had stalked and killed, all on my own, with a single well-placed bullet. I was elated when I ran up to the rabbit, and then I became profoundly distressed and saddened as I watched its death throes and its blankly staring eyes that saw nothing. I was keenly aware that I had willfully taken the life of this beautiful creature. I gutted the rabbit and put it gently into the game pocket of my hunting coat.

When I arrived at my granny's house, she saw the blood on my hands and noted the tears on grimy cheeks. She asked me brightly, "What's in your poke?" With some trepidation, I laid the rabbit on the porch railing. She picked it up, held it aloft by its hind legs, hefted it, and remarked on what a nice fat rabbit it was. She told me, "You go out back and skin him and cut him up for our supper." I did as she asked but without much enthusiasm.

That night she proudly announced to those gathered around the table that tonight's chicken-fried swamp rabbit was courtesy of the family's newest hunter. All seated around the table clapped and smiled their approval; my grandfather reached over and ruffled my hair. As the meal progressed, all were effusively complimentary of the chicken-fried swamp rabbit. That, of course, allowed me to spin an embellished rendition of the hunt. You would have thought this was the biggest, most delectable, most beautiful swamp rabbit ever killed in the entire history of mankind.

That rifle is still, to my mind, the most beautiful firearm ever made. I have treated it lovingly and respectfully for over seventy years now. The bluing is worn off many of the metal parts, and the varnish has slowly disappeared from the stock. But that old rifle remains in perfect working order and is without a speck of rust.

Uncle Henry died from "the cancer" when I was eleven years old. He left me his Remington single-shot rifle, a "brick" of 500 .22 caliber long-rifle cartridges, and an enormous education in the ways of wildlife as well as a love of hunting and fishing along with an array of hunting, trapping, and fishing skills—some acceptable and some not by today's standards.

When World War II ended and the nation's surviving warriors came home, our society's tacit approval of—or maybe it was tolerance for—subsistence hunting came to an abrupt end. As a result, I was required to temper my now well-honed hunting and trapping skills. Compliance with hunting and fishing laws was henceforth expected and required by my father and grandfather. It was made clear to me and my fishing and hunting buddies that noncompliance would result in our firearms being taken away and "wrapped around a tree." Even I could see that laws and regulations made more and more sense as I encountered ever more hunters in the woods and anglers along the river. It was a fork in my trail.

I had not, however, given up tossing homemade pipe bombs into the Trinity River to kill alligator gar and carp, which were, after all, known as "trash fish." Big Dad had acquired several drums of surplus military blasting powder for the purpose of blasting out stumps when clearing new fields. The Trinity River carried the residue of Fort

*Jack teaching his Aunt Margaret how to shoot his rifle,
Handley, Texas, 1952.*

Worth's somewhat ineffective sewage treatment efforts downstream to Dallas and on to the Gulf of Mexico. The joke in Fort Worth was "Flush twice—Dallas needs the water." Nobody in their right mind would eat whatever "game fish" lived in that reach of the Trinity River north of Big Dad's place.

One day in the summer of 1947, a buddy and I had just tossed a "low-grade" homemade pipe bomb from off a bridge into the Trinity River when a big man wearing cowboy boots and a big hat sauntered up. He had a badge on his shirt that proclaimed him to be a game warden. He greeted us in a friendly but authoritative baritone voice, "Howdy, boys. What y'all up to?"

I answered, "Howdy, mister. We're blowing up gars and carps. You want to throw one?"

The big fellow declined the invitation. He introduced himself. "Boys, I'm the local game warden. What's your names?"

"Howdy. I'm Jackie Ward Thomas and this here is Paul Moore." We had no idea of who or what a game warden might be.

We shook hands. The warden went on, "Jackie, is Scranton Thomas your dad?"

I nodded, "Yes, sir."

"And Mister Arthur Thomas is your grandpa?"

"Yes, sir, that's my Big Dad."

The game warden issued a pointed invitation. "Why don't you boys put your bikes and your possibles in the back of my pickup? I'll give you a ride to your grandpa's place."

When we pulled up, Big Dad and Dad were sitting on the front porch drinking iced sweet tea out of quart Mason jars. The warden told us to stay in his pickup and joined them on the porch. Pretty soon Dad waved Paul and me over. Fortuitously, he and the game warden had played football together at Handley High School in the late 1920s.

The warden told us that our use of explosives to reduce the Trinity River's population of "trash fish" was both illegal and dangerous. So maybe "it just might be a real good idea if we gave it up for Lent." Then he gave us a serious lecture. By its end, jail time, perhaps with torture thrown into the mix, seemed likely. Years later, Big Dad

Jack at Air Force ROTC summer camp, Williams Air Force Base in Mesa, Arizona, where he trained for six weeks in 1956.

confided that he, my dad, and the game warden were struggling to keep a straight face.

So my young colleagues and I gave up pipe bombs—then and forever. Over the next few years the game warden was to become something of a mentor for me. From time to time, he picked me up at Big Dad's place to ride along with him as he went about his patrol duties. I finally figured out that he was using me to show him hidden back roads and trails and shortcuts.

In the process, he talked to me about becoming a game warden or maybe a wildlife biologist—whatever that was—though the latter occupation required a college degree. I listened carefully, though college seemed highly unlikely for me, and I didn't have a clue as to what a wildlife biologist was or did. So his encouragement didn't make much of an impression at the time.

Six years later, in 1953, I graduated from Handley High School and headed off to the Agricultural and Mechanical College of Texas (now Texas A&M University) with the intention of becoming a veterinarian—I knew what a veterinarian was. I did okay in my course work as a freshman. Then in my second year I encountered organic chemistry, the only course I flunked in my entire college career. Though I passed with something close to flying colors the second time around, the more I thought about it, the less appealing doctoring horses and cows and dogs and cats seemed.

Jerry Don Cobb, a teammate on my high school football and basketball teams, was a year ahead of me at Texas A&M and was majoring in wildlife biology. I changed my major to wildlife management and never looked back.

CHAPTER 2

"NEW" CAN BE A VERY SCARY THING:
TEXAS GAME DEPARTMENT, 1957–1966

After graduating in 1957 from Texas A&M with a degree in wildlife management, I found gainful employment with what was then known as the Texas Game Department (TGD). My first assignment was as an assistant project leader for game management in the small town of Sonora in Sutton County, located where the western edge of the Edwards Plateau meets the Chihuahuan Desert. My first wildlife management district—Sutton, Edwards, and Crockett counties—was a perfect place at a most opportune time for a "newbie" wildlife biologist, arriving on the scene just when wildlife management was coming into its own, to learn and help invent what was to be my lifelong profession.

BUMP GATES AND BRUISED EGOS

After college graduation, I retreated to my parents' home in Handley for a few days before beginning my professional career on the first day of June. A letter from the Texas Game Department awaited, authorizing me to take possession of a pickup truck from a Dodge dealer in Arlington, seven miles down the road from Handley. A more senior biologist had left his old truck at that dealership when he took possession of a new pickup. The letter informed me that all the equipment I would need for my new job was in the keeping of that friendly Dodge dealer.

Two brand new game biologists for the Texas Game Department, Jack and Carl Schubert (right), 1958. The 1950 Dodge pickup served as their work vehicle.

My mom dropped me off at the dealership with all my earthly possessions: a duffel bag containing all my clothes, a 16-gauge double-barreled shotgun, a .22 caliber rifle, a pair of field boots, a rod and reel with a box of fishing tackle, a shaving kit, and several boxes of books. I had $50 in my pocket, a loan from my grandmother that would tide me over until a payday rolled around and I would start drawing my salary—$285 a month minus deductions.

The friendly Dodge dealer handed me a TGD property sheet to sign. All the items listed were contained in a large cardboard box and included a pair of 7×35 Bausch and Lomb binoculars, a well-used Royal manual typewriter, a box of wooden pencils, a black leatherette notebook filled with green unlined paper, a box of carbon paper, a gasoline credit card, a ream of typewriter paper, a roll of three-cent stamps, a box of envelopes, and assorted forms.

And, most important, there was a twenty-five-page operations manual that told me everything I needed to know about the functioning of the bureaucracy of which I was now a part. At the bottom of the box was a note written with a red felt-tipped pen: "WELCOME TO THE TEXAS GAME DEPARTMENT!"

I swallowed hard as I walked around my "vehicle" (government talk for any car or truck), a nearly eight-year-old half-ton 1950 Dodge pickup, battleship gray, with big round decals on the doors proclaiming "TEXAS GAME AND FISH COMMISSION." The fenders, all four, were rusted through, and the windshields on both sides of the center post were cracked. One of the door windows was stuck halfway up—or maybe halfway down. The boards of the truck bed were rotted through in several places. An old army blanket tied in place with surplus nylon parachute cord hid the places where the seat had worn through. The odometer read 27,000 miles, which was most surely 127,000 miles—or maybe even 227,000 miles.

I fired up the old truck, waved good-bye to the Dodge dealer, and rattled out of the parking lot with fenders flapping and exhaust spewing from the tailpipe. I was on my way to Sonora. I spent my first night on the job sleeping in the truck's cab in a roadside park just outside of Menard. I had no idea how to make a claim for a motel room—or even if such a thing was allowed. For sure, I couldn't afford to shell out three or four dollars cash money for a motel room.

When I rolled into Sonora, my first move, as instructed, was to look up Game Warden Nolan Johnson. In 1957 wildlife biologists were the new boys on the block. Game wardens were considered "Mr. TGD" in the counties that they patrolled. A few of the old-time wardens were less than enthralled with the "smartass" college-boy wildlife biologists being hired in increasing numbers and, in the view of some, invading their territories.

Fortunately, Nolan, a relatively recent product of training for game wardens at Texas A&M, and his beautiful blonde wife Helen seemed glad to see me. They took me into their home while I scouted around for housing for me and my soon-to-be bride (I was getting married in less than a month). I looked over all three places in the very small town that were available for rent. My choice was a freshly renovated apartment over a three-car garage in the backyard of the matriarch of the most prominent ranching family in Sutton County.

The rent, which included water and electricity, was $50 a month. I figured I could handle that on my munificent salary of $3,420 a year (before deductions for Social Security, state retirement, and health care). My soon-to-be bride, Miss Margaret Schindler, had landed a job teaching high school choir and kindergarten in the Sonora public schools at $3,600 for nine months' work. Margaret was too kind to make a point of difference in our work.

The landlady threw in the natural gas when she learned that my wife-to-be was a Methodist—and a music teacher—who might be recruited to fill the choir director vacancy at the Methodist church. To help things along with my landlady, I described my pending religious conversion from Presbyterian to Methodist. Margaret and I had decided this in a close one-to-one vote, which, as proved usual with such things, I lost.

During my first month on the job, I frequently rode shotgun for Warden Johnson as he patrolled for game law violations, visited with ranchers to maintain public relations (the land was all in private hands, and access to hunting was by sufferance of landowners), and, in general, "showing the TGD flag." Nolan, a World War II combat veteran of the Philippine campaigns, was short (about five feet nine inches), strong, and wiry, with a full head of curly black hair and coal-black eyes. He had a subtle sense of humor that included having fun at my expense.

Bump gates are wide, heavy, wooden gates that are mounted on a center pole about twelve feet tall. They are supported from the top of the pole by a twisted cable that holds the gate closed a foot above the ground. Gravity and the twisted cable allow a vehicle to push the gate open while simultaneously lifting it up. Once the vehicle is through the gate, the weight of the gate on the twisted cable allows it to drop and swing back into the closed position.

I had never even heard of a bump gate, much less seen one. So when we got to my first bump gate, Nolan stopped the patrol car. Before he could say or do anything, I jumped out and pushed the gate open, held it, and waved Nolan through. I shut the gate and hopped back in the car. Nolan smiled and nodded his thanks. Thereafter, at every bump gate, Nolan stopped and waited for me to hop out and open the gate. Then he drove through and waited for me to shut the gate and get back in the patrol car. This routine went on for over two weeks.

Then one day we approached a bump gate just as a pickup approached from the other direction and got to the gate first. Nolan stopped the patrol car, and we watched as the driver bumped the gate and drove through. The gate swung back and slammed into the gate posts on both sides, bounced once, and came to rest in the shut position. My mouth dropped open. I slumped in the seat, pulled my battered Stetson down over my eyes, and felt my face flush through several shades of red.

Nolan tried not to laugh but couldn't help himself. We both laughed until tears ran down our faces. He made me feel some better when he admitted that he had been raised in deep East Texas and he too had never seen or heard of a bump gate until he was assigned to the Texas Hill Country.

A veteran game warden to whom Nolan was apprenticed put him through the same rite of passage. He assured me that he would keep our secret—and burst out laughing all over again. He kept his word. I admit now that I put several rookies through the same initiation when given the opportunity.

During the two years that I lived in Sonora, Nolan and Helen were Margaret's and my best friends and valued mentors. He taught me, mostly through example, public relations skills that served me well over a professional lifetime. He guided me in fitting into a new subculture by understanding that respect has to go both ways. Over the rest of my career, I relied on lessons he taught me as I adjusted to the cultures of Appalachia, New England, the Pacific Northwest, the Northern Rockies, and Washington, D.C., not to mention India, Pakistan, Mexico, Poland, Scandinavia, Israel, South America, Canada, and Africa.

Over the next year, he introduced me to every landowner in Sutton County—and every other person who had any influence on, or even a mild interest in, wildlife and its management. He was invariably kind when he could have been indifferent. I never forgot the lessons he taught me about both people and wildlife. And I have never ceased to be grateful to Nolan Johnson.

Ten years later, when I was leaving the TGD to join up with the USFS in West Virginia, I called Nolan to thank him for what he had done for me in those early days. I told him that if there was anything I could ever do for him, he had only to ask. He replied that when opportunity arose, I should simply "pass it on." I knew good wisdom when I heard it. In every job assignment afterward, I tried to do just that. My professors at Texas A&M provided gems of wisdom that served their students well. But it was Nolan Johnson who put those lessons into context for me.

First, wildlife management is mostly people management. Second, the most sensitive nerve in the human body runs between the heart and the pocketbook. Third, all things are political, and all politics are local. Fourth, everything is connected to everything else, and there is no free lunch. Wildlife managers had to develop the people skills to give politics and money their due role in natural resources management. The key to success was intertwining and balancing those four truths, always keeping the welfare and proper management of wildlife in mind.

Neither I nor any of my colleagues really knew what our profession and our careers would evolve to entail. We were, by and large, making things up as we went along and in the process, whether we knew it or not, helping invent our own profession. It was a big help that most of the game wardens treated wildlife biologists with kindness. I was, am, and will always be grateful to those good men and my esteemed colleagues.

After some six months on the job, I was told to bring my old Dodge truck to Austin, the state capital, and pick up a brand-spanking-new 1957 half-ton Chevrolet six-cylinder pickup. My new truck was a beauty, but with a few oddities. For example, per agency policy, the AM radio had been removed from the dashboard (at an added cost by the dealer) lest such a "luxury" item indicate to taxpayers that we were being less than frugal with their tax money. By the same reasoning, state vehicles had no air conditioning, although temperatures could reach above 100 degrees for more than six months of the year. But, all in all, I was proud of and grateful for my brand-new truck.

I did feel a twinge of sadness walking away from my old Dodge, which was, at long last, destined for the junkyard. Even today, I sometimes have dreams that involve that old truck. I can say this: the people of Texas got their money's worth.

EXTIRPATED? NEVER SAY NEVER

During my second year of drawing a paycheck as a genuine professional wildlife biologist, I was charged with operating a check station where hunters who had killed antlerless deer (usually does and fawns of both sexes) were required to appear with their kills. That check station was located in Edwards County, just east a mile or so outside of the county seat of Rocksprings, at a wide spot beside a road where gravel and crushed limestone were routinely stockpiled. Clouds of caliche dust rose in clouds when a vehicle entered or left the area.

The hunting season for antlerless deer was the last two weeks of a forty-five-day deer season at the end of the calendar year. The hours were 8 A.M. until 9 P.M., a long, largely uneventful, thirteen-hour day for anyone manning one of those check stations. In 1959 hunting for antlerless deer was a first-time event for Edwards County and had not yet been widely accepted by most landowners and hunters.

In the early afternoon of a sunny day in the last week of December, an old Ford pickup, with rusted-through fenders rattling, slid to a stop in a billowing cloud of dust next to my TGD pickup. I was seated on the tailgate "just spitting and whittling" to pass the time. As my level of irritation settled along with the cloud of caliche dust, the driver dismounted, slammed his pickup door, and walked toward me. I was not a happy camper but met him halfway with an extended hand and something of a forced smile.

The grizzled old-timer had a wad of Day's Work chewing tobacco stuffed in his cheek. The color of his remaining teeth gave evidence of his lifelong habit of dipping snoose and chewing tobacco. He leaned over and spat a stream of tobacco juice and wiped his mouth with the back of his hand. "Are you the TGD biologist?" The word "biologist" came out with an intonation that led me to believe that no respect was implied.

I stood up. "Yes, sir, my name is Jack Thomas. Can I help you?" I extended my hand.

He shook my hand with a stronger grip than was warranted—I could feel the calluses. "Son, I want to know just one damned thing. What's y'all gonna do about the bear that's gettin' into my angora goats?" It was more challenge—maybe a demand—than an inquiry.

I was a bit taken aback. "Sir, black bears were extirpated in this area somewhere around about 1910."

He spat again; the impact in the dust was getting closer to my boots. His eyes were flashing black under bushy, dusty gray eyebrows. He stepped closer. I didn't step back. "Okay, you government smartass, just what the hell does 'extirpated' mean?"

It seemed wise to take my hands out of my pockets, raise them waist high, and take a deep breath. "Well, sir, 'extirpated' means that a species is locally extinct."

The old rancher squinted. "Species? Locally extinct?"

I struggled to be respectful while extracting my size 12 boot from my mouth. "Well, sir, 'extirpated' means 'local extinction.' In other words, there are a lot of black bears in the United States—even a few in East Texas, south of us in Mexico, and in far West Texas. But because black bears were killed out in this part of Texas by the early 1900s, we say that they are 'extirpated'—they were here once but not now."

The old rancher turned his head and relieved himself of another mouthful of tobacco juice—even closer to my boots. I was beginning to believe that there was some message in this behavior. He wiped his mouth with the frayed sleeve of his faded denim jacket. "So, let me get this straight. Y'all are telling me that it ain't no bears gettin' into my goats?"

"Well, sir, I'm not saying you don't have a predator problem, but it's highly unlikely that it would be a bear. I'll call the local game warden, Ellis Martin, on my two-way radio and ask him to be in touch with you. He can arrange with the TGD trapper to give you some help."

I wrote down the fellow's name and telephone number. He assured me that Ellis knew him and where he lived. And there was no use in anybody trying to call, as the phone line from the county road to his house was down and had been for a couple of weeks. With that, my new friend spat again, about six inches ahead of the toe of my boot, and walked to his truck. His scuffed boots, spurs attached, raised puffs of dust with each step. His old Ford pickup started up after several grinding attempts. He was off—literally in a significant cloud of dust. It seemed to me that he spun the tires to provide emphasis to his disdain.

I dug around in the gear box in the bed of my ancient truck for my copy of *Hunting, Trapping, and Fishing Laws of the State of Texas.* The section related to black bears, obviously enacted long after extirpation had occurred, prohibited, at the risk of significant penalties, any attempt by any means by any person to take, kill, harm, harass, etc., etc., a hair upon the head, or ass, of any black bear that might exist in the western two-thirds of the sovereign state of Texas.

Later in the afternoon, the game warden showed up. He poured himself a cup of the cowboy coffee I had simmering on the Coleman stove on the tailgate of my pickup. I passed on the report that the old rancher had given me.

Ellis saw some humor in my explanation of "extirpation" to the old-timer. He cautioned me that such fancy language was one of the reasons some locals thought wildlife biologists were a little strange— and a bit on the smartass side. He smiled as he asked me if I knew what "extirpation" meant before I "done went to college." I blushed and admitted that I didn't. He promised he would call on the rancher and, if necessary, arrange for a TGD trapper to give him some help.

Three days later, the rancher's old truck roared into the check station, a tad too fast and raising a huge cloud of dust. I interpreted that as some sort of statement. I had just finished collecting required data from a hunter and certifying his antlerless deer as legally taken. That hunter and his partner were toasting their backsides at my fire of mesquite logs. With some trepidation, I walked through the settling cloud of dust to greet the driver.

The rancher had walked around to the back of his truck and dropped the tailgate with a bang. Just as I walked up, he reached into the truck's bed and tugged something toward him. I heard a heavy thud and saw a plume of dust rise as a very dead yearling black bear hit the ground.

I stood there slack-jawed. With a flourish, the old-timer slammed the tailgate back into its upright position and latched the chains. The ensuing dramatic pause allowed him time to relish the stunned look on my face. Then he drew himself upright and spat just in front of the toes of my boots—the closest yet! That flourish, how-ever well executed, was beginning to irritate me.

"Don't worry, sonny boy, it ain't what you think. I have to ad-mit the sonofabitch fooled me too at first. I thought for a second there, it for damned certain was a gawd-damned bear. But then I thought, 'Naw, that couldn't be! After all, black bears have been extirpated from these parts for nigh onto sixty years. Must be true—the biolo-gist told me so. Heck, everybody knows how smart those guys are!'"

He climbed back into the old truck, slammed the door, ground the engine to a reluctant start, and spun the tires in the caliche dust all the way to the pavement. He had his hand out the window waving gleefully as he drove away. I guess I should have been grateful that he wasn't waving his middle finger. I figured I would hear this story for years to come.

Right then, I pledged to never ever say "never"—and sure as hell not to say "extirpated" except to another wildlife biologist. Lesson learned.

LIPSTICK ON A PIG

When I started work with the TGD, I had only a vague idea of what a wildlife biologist was—or did. So logically enough, I "imprinted" on my boss, James Garth "Jim" Teer much as a newly hatched gosling imprints on the first moving thing it sees after hatching, usually its momma. Over time, by sticking close and paying attention, the gosling learns what a goose is and how a goose is supposed to behave.

I had arrived at a fork in the trail, and it fell to Jim Teer to point the way. Even today, I still wonder from time to time just how my professional life would have fared if my first boss and mentor had been less curious, dedicated, patient, ambitious, inclusive, well educated, bright, demanding, and sharing than Jim Teer.

Jim was a tight man with a dollar—or a dime, for that matter. That frugality extended to expenditures on his meager wardrobe. He was saving every nickel to go back to school at the University of Wisconsin to finish up his work on his doctorate.

Jim's office was on the second floor of the Llano County Courthouse—the door opened onto the courtroom. Two Texas Highway Patrol officers had a desk in the same office they used to fill out their reports. Officer William "Bill" Shipp was a big man, loud, boisterous, bright, and he picked and prodded at any available target for entertainment. Teer, who was even brighter than Shipp, was considerably more on the reserved side and therefore a frequent target of Shipp's disparagements.

It was not unheard of for a game of dominoes to break out in that office during lunch hour or after the close of official business hours. When such occurred, the office door slammed shut, a tabletop was cleared, and the dominoes rattled in the first shuffle. Shipp's style was to subject his opponents to an unceasing string of slurs, pokes, jabs, and veiled insults. Why? As a distraction, maybe? Usually some insignificant monetary wager was involved, though Teer considered anything in excess of a nickel or a dime significant; a quarter was highly significant.

The burly highway patrolman was the ultimate competitor and the world's poorest loser, no matter what the competitive activity, which ranged from dominoes to poker to tennis to fishing to hunting to "spitting for distance." Any competition was a serious matter to Officer Shipp. When he was losing, he offered a string of side bets—the more bizarre and distracting, the better—as a chance to get even.

One noonday, Shipp walked into Jim's office, slammed the door, cleared a tabletop, and challenged Jim to a game of straight dominoes. Jim went on a roll, winning game after game. When it was time to go back to work, Jim was counting up his winnings. Shipp slapped down a two-dollar bill (which was worth a whole lot more then than now) and proclaimed, "I'll bet any man here that Teer has a hole in both of his socks—*and* a hole in the ass of his drawers!" All looked puzzled. Where the hell did that come from?

Jim, blushing, teased the money out of his wallet and slapped it down on the table. Game on! Jim propped up a foot on the desk, unlaced a battered Red Wing field boot, and tugged it off. A toe showed through the blue sock. Shipp pumped his fist, "Yeah!" Jim unlaced his other boot and pulled it off. A white sock emerged. The toe was intact, but there was a hole—very small but a sure-enough hole—in the heel. Pressure was building. Onlookers began to clap in unison and chant, "Take it off! Take it off!"

Jim's face flushed. Dead game, he stood up and dropped his jeans to his knees. There was a barely discernible hole on the left-ass side of his Jockey shorts. The patrolman snatched up the money, stuffed a fresh cigar in his mouth, opened the office door, took a parting bow, and clattered off down the steep metal-clad stairs.

Now fast-forward a quarter century to the North American Wildlife and Natural Resources Conference in Washington, D.C., in 1984. By then, I was the chief research wildlife biologist for the USFS, and I happened to run into Dr. James G. Teer, who was now head of the Department of Wildlife Sciences at Texas A&M University.

We were both, by the standards of our profession, nattily dressed in neatly pressed suits and ties. Jim was accompanied by a bevy of fresh-faced, clearly admiring graduate students, and he graciously and formally introduced me to his entourage. I introduced Jim to my USFS associates. It seemed a very long way—in both in time and circumstances—from our days working and playing dominoes together in the Llano County Courthouse.

It dawned on me that the budding wildlife biologists surrounding Dr. Teer could use a dose of reality—and humility. So I told them the story—only somewhat adorned and exaggerated—about "holes in both socks and the ass of his drawers." Jim was tolerant, having no real alternative. The admiring graduate students, not knowing what to think, seemed a bit befuddled. "Ladies and gentlemen," I said, "remember this: you can put lipstick and a dash of perfume on a pig, but it's still going to be a pig."

They got my point. It's good to remember that we really don't change all that much as the years pass. Titles, matching socks, shorts without holes in the ass, and a little more cash in the pocket don't make the person. And Dr. James G. Teer had always been and would always be the kind of man I aspired to be, in spite of a hole in his shorts at a critical moment.

Throughout my career, I remained, down deep, the same slightly socially awkward introverted nerd from Handley, Texas. I had simply learned to fake it as times and circumstances evolved. I often wondered, "What the heck am I doing here?" and "What if they figure out that I'm not really who they think I am?" I was pretty sure Dr. Teer had some of those same feelings. When I had an opportunity, as part of my duties, to shake hands or consult with several presidents of the United States, I wondered if they had those same feelings—and I thought it likely. After all, a pig is still a pig in spite of all the lipstick and perfume (university degrees, big job titles, awards of one kind or another, and so on). One learns to smile and live with it.

ON THE SHOULDERS OF OTHERS

When I started work for the TGD in June of 1957, only a very few of the wildlife biologists in the agency had one or more decades of experience in research and management activities such as trapping and transplanting quail, deer, antelope, and turkeys into vacant habitats. Biologists were just getting into the management of hunting and fishing as one county after another chose to be included under TGD's regulatory authority—the setting of all rules and regulations for hunting, fishing, and trapping—through a commission appointed by the governor. More and more wildlife and fisheries biologists were being hired to execute those blossoming programs; in fact, the agency hired the entire Texas A&M graduating class in wildlife biology in 1957. In contrast, game wardens had been employed in steadily increasing numbers since the early 1900s.

By 1957, nearly every one of the 254 counties in Texas had a resident game warden, many with more than a decade of experience. Wardens commonly knew most if not all of the landowners in their assigned work areas, along with most of the other citizens with any political clout or interests relative to fish and wildlife matters. Public relations were a significant part of their job description—perhaps the most significant part.

Many, if not most, wardens spent as much time protecting landowners against trespassers as they did enforcing other regulations. That focus, understandably enough, increasingly endeared them to landowners as the sale of hunting rights became a more and more significant means of income.

Under the rhetoric of the much-praised North American Model of Wildlife Conservation, wildlife belongs to the people as a whole, while access to that wildlife is at the discretion of the landowner. That discretion allowed landowners to charge would-be hunters for access to their property for purposes of hunting. In Texas and elsewhere, more and more landowners, under the guise of charging for access, began to charge hunters by the head harvested. At the very least, that seemed to me to be a perversion of the vaunted conservation model.

But it proved to be a very successful fiscal perversion that has allowed landowners and hunting to prosper from selling hunting privileges to hunters that could and would pay the associated fees. That produced increases in wildlife populations, numbers of hunters, and wildlife legally harvested as well as more prosperous ranching enterprises.

The only losers were those that Aldo Leopold, the father of wildlife management in North America, referred to as "one-gallus hunters" (those who were so poor that they could only afford one suspender to hold up their pants). Clearly, it was an increasingly marked deviation from the North American Model of Wildlife Conservation, which entrusted ownership of wildlife and its management to the states except for migratory wildlife and species declared to be threatened or endangered under federal law, which were under the auspices of the federal government.

Until the 1950s, game laws and regulations in Texas had been determined, county by county, by the state legislature and thus varied widely, sometimes dramatically. Often game wardens with finely honed political skills arranged to have laws enacted that they, as individuals, considered important. The laws related to fish and wildlife for a single county could take up page after page in the book of laws. Results were sometimes ludicrous, such as having different fishing regulations for counties on opposite banks of the same river.

In short, old-time game wardens had been the alpha and omega of matters related to fish and wildlife in the counties of their assigned territories. That status had been hard won. A work week commonly entailed fifty to sixty hours on duty—and even more during hunting seasons. Working in remote areas, usually alone and often at night and on weekends and holidays, occupied most of that time.

In the 1950s, regulatory authority began passing, county by county, to the Texas Game and Fish Commission. Counties were grouped into fish and wildlife management units on the basis of geographic and ecological similarities. Rules and regulations, formulated by supervisory fish and wildlife biologists, were simplified and made uniform within each management unit. Those recommendations were formalized and adopted by the commissioners, sometimes with adjustments.

The younger game wardens, who were graduates of an intensive six-month training course at Texas A&M, understood the role of wildlife and fisheries biologists. These new-school wardens expected to work closely with biologists. Many old-school wardens welcomed, or at least tolerated, the evolving situation, but a few viewed the "newbie" wildlife biologists as a potential threat to their stature and authority.

Many old-line wardens "kept score" on how well they performed their jobs by the number of convictions for violations of game laws. Since the old regulations often contained various confusing and illogical regulations that could be unintentionally violated by hunters and fishers, they gave the game wardens so inclined more chances to rack up a score.

A primary duty of wildlife biologists in many areas of the state was dealing with dramatic overpopulations of white-tailed deer that were damaging their own habitats. The recovery of deer populations from the lows of the early 1900s had been slow, difficult, and sometimes dangerous for the warden force. Much of the credit for success went, and rightfully so, to game wardens who established rapport and *pro forma* partnerships between landowners and the TGD. The dramatic recovery in deer numbers initially resulted from limiting hunters to taking only branch-antlered bucks (generally, bucks over two years old) with absolute protection for the spike-antlered bucks (usually less than two years old), does, and fawns.

As time passed, ranchers garnered rapidly increasing revenues from leasing hunting rights. As might be expected, the more money they made from selling hunting rights without damage to their resident deer herds, the greater their concerns with the welfare and numbers of game species became. To protect both livestock and wildlife, predators like wolves, coyotes, cougars, eagles, and bears were relentlessly pursued and dramatically reduced in numbers. Some predators, such as wolves, coyotes, and cougars, were extirpated over most of Texas by the 1950s.

Deer, turkey, and antelope populations surged in the 1940s and 1950s. A new day was dawning for wildlife management as more and more wildlife biologists were employed; indeed, the "times they were a-changin'." Now a few old-school game wardens were increasingly uneasy—and a very few somewhat hostile—to the "invasion" of their territories by young college-trained wildlife and fisheries biologists.

A few biologists reacted with undue glorification of their own credentials, which some wardens interpreted as disparaging of the expertise and importance of the warden force. In some cases, at least for a time, there was a "failure to communicate" between game wardens and biologists.

Sticking points in warden/biologist relationships sometimes, though rarely, came out in annual public hearings. Biologists reported on the status of wildlife surveys, explained changes in hunting and fishing regulations, and discussed the likelihood of issuing "doe tags" to landowners to enhance hunting opportunities (and, potentially, income to landowners).

Some landowners—called "early adopters" by sociologists—smelled the money and eagerly accepted the new program of harvesting a regulated number of antlerless deer. Some, remembering the long struggle to rebuild game numbers, were more hesitant. Now a new day had dawned with dramatically new circumstances. The bastardized term "unselling" best described efforts to overturn long-standing, highly successful campaigns against killing female deer and to include antlerless deer in carefully regulated harvests.

For me, that tension came to a head in a public hearing when an old-school warden took the floor and repeated a popular myth about whitetail behavior. Half of the audience nodded their heads in agreement. Others, better informed, shook their heads in disagreement and disgust. The warden had put me on the hot seat—and, I thought, did it on purpose. If I spoke up, I would insult him and his admirers. If I said nothing, those who knew the facts would consider me weak-kneed. I sensed a fork in the trail.

I stood and carefully, thoroughly, and (I thought) respectfully laid out the facts as I saw them. What I said was, in my opinion, irrefutably true, clearly presented, and essentially impossible to deny.

But judging from the reactions of some in the audience, it was an ill-advised political move on my part, wrong in time, tone, audience, and place—wrong all the way around. The old warden's face flushed as he crossed his arms over his chest, jutted his jaw, slid down in his chair, and glared.

As the courtroom emptied, the clearly angry warden walked up to me. My hackles came up. I raised my hands higher than my hips just in case he took a shot at my nose. Then he totally disarmed me.

"Kid, I hate to admit it, but you and your biologist buddy know your business. It's just really hard for us old-timers to give up being 'Mr. TGD' in our counties. Deep in my heart, I wish you biologists—and regulatory authority—would just go away and let things go back to where they used to be. Those were the good old days for me. But that's not going to happen, and I know it shouldn't.

Times and circumstances change, and I have to change with the times. That's hard."

Then he added, smiling, "But I want you damn *bugologists* to remember something. We old-time wardens are the ones who held the line for wildlife and helped turn things around while you were hanging on to your momma's skirts. For damned sure, it wasn't easy. You young biologists have no idea how hard and how dangerous that was. A few of us died getting it done. Whether or not you want to admit it, you youngsters stand on our shoulders. You need to own up to that when you stand up to talk in front of our folks. Let's make a deal to show each other respect in public and present a united front. Doing what we get paid for is more important than egos, feelings, and minor-league pissing matches."

He was right, honest, forthright, and gutsy, a class act from start to finish. I knew him as a very good man. I had been angry enough with him, off and on, over the previous couple of years to consider throttling him. Obviously, he felt the same way about me. But just then we stood in each other's boots. I could see things his way and felt very much the arrogant youngster. I nodded and extended my hand. He took it.

We "young whippersnapper" game biologists in the 1950s and 1960s really did stand on the shoulders of the game warden force, especially the old-timers. It was only fair that we acknowledged and honored that debt. From that moment on, I went out of my way to do just that. If you want respect, accord respect. Or, as one old warden put it, "What goes around comes around."

Over the course of a half century in the business of conservation, I slowly morphed from a young whippersnapper into an old-school wildlife biologist. Now I cringe when new issues in conservation arise and I see the contributions and valiant efforts of my generation of natural resource managers denigrated. Like the old-school game wardens in Texas so long ago, we did the very best we could under the circumstances of the time. Really, what more can anyone do?

Those who live long enough and are sensitive enough to gain perspective should acknowledge that we stand tall on the shoulders of those who came before. Those who come after us will inevitably do the same. And so it will ever be. It's just too bad that such an obvious, simple lesson has to be learned over and over again.

THE ONLY GAME IN TOWN

My dad visited Margaret and me in Sonora for a week in the late spring of 1958, his first visit to our new home since our marriage in June of 1957. During his visit, as the new TGD wildlife biologist, I conducted a solo county-wide public meeting on proposed hunting regulations, including the likelihood of an open season for legally taking antlerless deer in that county for the first time in over a half century.

Many of those seated in the courtroom at the county courthouse had kept their big hats on; with their arms crossed and their grim expressions, they were hard to miss. I wasn't far into my presentation before it dawned on me that I had walked into something of an ambush set up by the county's veteran game warden who, as he put it, "didn't cotton much to doe killing." He didn't hide his resentment of a young wildlife biologist who had come into "his county" proposing changes with which he did not agree.

The lean, gray-haired, old-time warden sat in the front row in the hearing room with his Stetson pulled down to shade his eyes, arms and legs crossed, and a defiant look on his face. I struggled through the meeting. Having my dad watching, with his jaw set and his blue eyes increasingly flashing with anger, made things worse.

I held my temper, even as I felt my face alternate between flushed and pale over the course of the meeting. When Dad and I got back to the camp house, I vented. Comments such as "dirty politics," "underhanded sonofabitch," and "sneaky bastard" spilled out. Dad listened patiently until I ran out of gas.

Dad calmly asked, "Is what you're trying to do important?"

"Damn right!"

Dad kept probing. "And you think what happened tonight was unfair?"

I nodded vigorously.

"It was dirty pool, right?"

"Damn right!"

Dad lifted a bushy black eyebrow flecked with gray and asked, "Why are you surprised? The guy's ego was at stake—and a bunch of money for the ranchers. They *know* him. They respect and trust him, probably for good reasons. And most of them don't know you from Adam. Let's face it, you're barely old enough to vote."

My father often taught me lessons through stories. I never knew, and still don't, if he made up his "teaching stories," but they made his points clearly and memorably, which compelled me to listen. Dad began his lesson. "When I was about sixteen, a group of young guys sometimes gathered under the streetlight in the alley behind the corner drugstore in Handley to shoot craps. On one such occasion, my second cousin—'Cuz' (a deputy sheriff)—busted the game. He stormed up, hog-leg six-gun pointed toward the sky, yelling 'Y'all are under arrest! Nobody move!'

"Somebody smashed the low-hanging streetlight, and it got pretty damn dark pretty damn quick, as everybody scattered. Two of us got tangled in our collective feet and fell down. Cuz jumped on top of the tangle. When Cuz got his flashlight into operation, he was sitting on me, his very own blood cousin. He looked at me and glared. I looked at him and smiled. Now, how was Cuz going to tell his Aunt Pearl that he had arrested her pride and joy for shooting craps—in a back alley at that? Surely, he feared that Aunt Pearl would eat him alive and sic her sister, his very own momma, on his happy ass. He knew it. I knew it. We both knew it.

"But he gave me a lecture and some advice. What the hell are you doing in that crap game? First, it's dumb to gamble. Second, you've got to know the game's probably crooked! He loosened his grip on the bib of my overalls and stood up. As soon as I could breathe a bit more normally, I replied, 'Cuz, we both know the game might be crooked. But I love to shoot dice, and it's the only game in town.' With that, I took off running.

"Now, son, think on it. This wildlife management stuff is just another game. Like all games involving money, ego, and politics, it's apt to be crooked to some degree. You represent the government, so you have to play by the rules. But that doesn't mean you can't play smart. Think chess, where you have to try to think several moves ahead.

"You assumed you were in the right and the facts were on your side. Assumptions can be dangerous. Your Big Dad used to say that 'assume'—taken apart—means to make an 'ass' out of 'u' and 'me.' These folks don't really know or care much who you are. You assumed folks would just buy what you deem your superior knowledge and not question your motives. You didn't explain what you were trying to do or how or why. You didn't understand that you were challenging that old-time game warden on his turf. You failed to accord him the respect he damn well earned and deserved.

"The local folks know him, and have for twenty years or so. They trust him. They don't know you—not yet. When you didn't show proper humility and respect to the old warden and to the folks in attendance, they didn't appreciate your attitude. They interpreted your presentation as an attack on them. You need to cultivate, explain, solicit opinions, be deferential, give credit—but keep your eye on the prize. The end result is what counts. It may take some time to get there, longer than you would like. But getting there is what counts the most in the end.

"The game is crooked; all political games are, to some extent. But it's the only game in town. You don't have to cheat—and you shouldn't—but you do have to play smart."

I took Dad's advice to heart and worked constantly at developing and honing my public relations skills in dealing with game wardens, ranchers, hunters, politicians, bureaucrats, and everyday citizens.

Twenty-nine years later, as chief of the USFS, I frequently sat at witness tables in front of Senate and House committees. When the going was tough and was sometimes, in my opinion, outrageously unfair, I took a deep breath and remembered my dad's words, "Yes, the game is crooked and the deck is stacked, but it's *the only game in town*. As an honorable man, you don't cheat, but you play smart and think several moves ahead. And never ever let the bastards get your goat."

Over the years, in a number of roles and sets of circumstances, I got better and better at following his advice, which proved, as always, good wisdom.

Many decades later, during a hearing before a hostile Senate committee in Washington, a senator was in the middle of what I considered a bombastic tirade that was clearly intended to intimidate me when he detected a slight smile on my face. He was hugely insulted, or pretended to be. He sneeringly demanded, "What are you smiling about?"

"I'm sorry, sir. I didn't mean to smile. If I did, I meant no disrespect. Please continue." In fact, I was visualizing my wizened bombastic protagonist in his birthday suit, an old trick that Big Dad taught me to tamp down fear and resist intimidation. It was, after all, the only game in town—although one far removed from that courtroom hearing in Texas so long ago. Now I was a player in a much bigger arena, a somewhat good one and getting better every day.

RATIONING THE B.S.

One of the first public meetings I conducted on my own hook as a "newbie" biologist with the TGD took place in Rocksprings in Edwards County in the fall of 1959. I had worked hard on my presentation using every trick I could remember from my public speaking classes at Handley High School and Texas A&M University.

I had carefully written out my presentation, read it over at least a dozen times, edited the draft each time through, and transferred the whole presentation to notes on three-by-five-inch file cards. I practiced until I could deliver my spiel without even glancing at the cue cards, but they were in my shirt pocket if I were to falter. In addition, I had prepared charts and graphs on posterboards, hand-lettered with crayons, to illustrate my key points.

I arrived a half hour early to size up the room, set up my visual aids, and read over my cue cards just one more time. When the advertised starting time came around, Ellis Martin, the local game warden, and I were the only folks in attendance.

After a half hour, I was gathering up my materials and preparing to leave when an elderly rancher ambled in. Ellis introduced us. The rancher took a seat in the middle of the front row bench in the courtroom and looked at me expectantly. I launched my presentation.

My Presbyterian Sunday school teacher had taught me long ago that the Good Book said that the size of the congregation didn't matter—the message was everything. The grizzled old rancher paid rapt attention and clapped his calloused hands vigorously at the finish. I opened the floor for questions.

The old rancher said, "Sonny, I don't have any questions. But I'm mighty thirsty and a little hungry. Would you boys join me for a chicken-fried steak and a cold Lone Star beer?" Though the game warden had another appointment, or so he professed, I was pleased to accept. I had built up a mighty thirst, and I was more than a little hungry to boot.

The only cafe in Rocksprings was just across the street from the courthouse on the town square. A neon sign in the window flashed off and on announcing "EATS" in four different colors. When we stepped out of the blistering sun into the relative darkness of the cafe, the cool, damp air coming from a swamp cooler mounted in the back wall was most welcome. I took off my shirt with the TGD patch on the shoulder, turned it inside out, and hung it on the back of a chair.

We state employees were expected to keep our "sins"—most certainly including the consumption of alcohol—private. After all, somewhere in the shadows, there just might have been a hard-shell Baptist lurking and eager to uncover inappropriate behavior by a state employee.

My T-shirt was damp with sweat. The beer bottles that the tall blonde buxom waitress set on the table were so near freezing that the beer contained a few ice crystals. Humidity from the swamp cooler condensed on the long-necked bottles and ran down to puddle on the cheap red-and-white checkered oilcloth that covered the table.

We eagerly downed a couple of "lifesavers" before being served our nearly overflowing platters of chicken-fried steak smothered in white-flour gravy with ranch-style beans, coleslaw, and corn bread on the side. I was hungry and it went down easy. After a dessert of peach cobbler, topped with a couple of scoops of vanilla ice cream, the grizzled rancher rocked his chair back on two legs.

He pushed his greasy Stetson back on his head and rolled himself a Bull Durham cigarette. He offered me the makings. I held up my hand in polite refusal: I hadn't acquired the knack of rolling a cigarette. Besides, Granger pipe tobacco was cheaper. I thought that, at age twenty-three, a pipe made me seem even more sophisticated and mature than I actually was. I packed my corncob pipe, and we both lit up.

The kindly old rancher had taken a liking to me, or maybe he was just showing a little Christian charity to a demonstrably ill-at-ease greenhorn. He leaned forward on the table and asked, "Son, are you open to advice from your elders?"

I leaned forward, looked him in the eye, and nodded, "Yes, sir."

He leaned forward and looked me in the eye, "Can you handle straight talk?"

I leaned forward and put my elbows on the table and looked him right back in the eye. "Yes, sir!" He took in a deep draw on his cigarette. I puffed on my pipe. It was quiet for a pregnant moment.

He began. "Well then, let me tell you a story my old daddy told me. He was working up in Wyoming hauling hay to wintering cattle. He got word from the fence rider about a traveling preacher man who would hold services at the nearest crossroads on Sunday next round about noon.

"It was mighty lonely for my old daddy with nobody or nothing but horses and cattle to talk to for days at a time. So he figured, come Sunday, to drive the hay wagon the hour and a half to the crossroads. He looked forward to listening to the preacher. The simple sound of a human voice, even that of a preacher man, would be welcome. Some of them circuit riders could do some mighty fancy preaching.

"So, come Sunday, he got up plenty early, fed the cows, and crawled up in the hay wagon, loaded with his bedroll, a tarp, and a couple bales of hay for bedding or food for the team in case the weather turned nasty. He popped his whip and he was off for the crossroads. He got there a bit more than an hour before preaching time, built a big stand-around fire, cooked up some coffee, and put his Dutch oven, with already-cooked beans and bacon, next to the fire to warm.

"About noontime, he saw a lone rider on a mule coming from the direction of Cheyenne. Sure enough, it was the preacher man, a young fellow likely just out of preaching school. He invited the pilgrim to dismount and warm by the fire. He seemed grateful for the coffee offered. When preaching time came, there was not another soul in sight. My old daddy allowed as how, if the preacher agreed, they could just eat some beans and corn bread and forgo the preaching.

"The preacher opened his worn Bible and quoted from the book of Matthew: 'Where two or more are gathered together in my name, there I am in the midst of them.' He got up in the bed of the wagon, stood up straight, and cut loose with a full hour of well-honed hellfire and brimstone.

"My old daddy just stood there bareheaded and listened. His nose was snotty and dripping, and his ears were stinging from the wind and cold, but no more so than the preacher man's. At the end of the sermon, his 'Amen!' was loud and heartfelt. He invited the preacher for a bit of hot coffee, beans, and corn bread.

"As they nursed the last of the coffee in their tin cups, the preacher man got around to asking my old daddy what he thought of the sermon. Daddy stirred the dying embers of the fire with a stick and cogitated on his answer.

"Then he looked the preacher in the eye. 'Preacher,' he said, 'when I'm out to feed the cattle and just one old scraggly cow shows up, I feed her and feed her well. But I sure as hell don't give her the whole damn wagon load.'"

The old rancher searched for understanding in my eyes. "Son, when just one or two pilgrims show up at your wagon, don't give 'em the whole load."

That old rancher could deliver a pretty good short load himself. Since that time, I have sized up my audiences, in both numbers and composition, and adjusted "the load"—in terms of length, content, and complexity—to fit the circumstances, and I've gotten passing good grades at public speaking in the process. How good? Good enough to get paid for it, and that's as good as it gets.

HE'S OUR SONOFABITCH

By the late 1950s, with the explosion of white-tailed deer numbers in Texas, the deer in some places were what might be legitimately called superabundant—to the point of depleting their habitats. Competition with livestock, especially sheep and goats, was obvious, and habitat was being increasingly damaged. Shrubs and trees were being "high-lined," with no leaves or twigs left as high as a deer could reach standing on its hind legs.

Still, killing does as a means of increasing harvest and controlling deer numbers was viewed by many hunters and landowners as a return to the bad old days. However, more and more landowners in the Texas Hill Country were making money at steadily increasing rates from leasing hunting rights for buck deer, which in some cases equaled or exceeded that from raising livestock.

Because money was involved—relatively "big money" at that— the politics surrounding doe killing was intense. The politicians, such as governors, legislators, and county commissioners, had learned that messing with hunting and fishing regulations was something of a "third rail" in Texas politics, a no-win situation: if you touched that electrified third rail, you lost elections.

The setting of hunting and fishing regulations was being increasingly turned over to the TGD, whose employees made recommendations to the agency's politically appointed commissioners for final decisions. In Texican, such games were delicately called "putting the turd in somebody else's pocket."

At that time, wildlife biologists were comparatively few in number. We conducted annual surveys to get a handle on deer numbers, distribution, sex ratios, and reproductive performance. Armed with that data, we held annual public hearings in each county to discuss our findings and recommendations. Those attending the hearings were welcome to voice their opinions and ideas, which were forwarded in a report to the commissioners.

Some powerful landowners, usually those with the most land and political clout, were accustomed to bringing legislators around to their way of thinking through a variety of means. They were now facing off with a bevy of young, somewhat idealistic wildlife biologists who were just learning the ways of the game, technically and politically. Most wildlife biologists were trained—whether we realized it or not—in the philosophy of the Progressive Era and were, by training and inclination, long on "science" and "scientific management" and way short on political acumen, which would come only with "on-the-job training." That training ensued upon employment and continued without respite throughout a career.

During the 1950s, when the TGD first began to exercise regulatory authority in a few counties, field biologists worked diligently to become acquainted with landowners in their districts. Most of the young biologists were not too sure of themselves and therefore operated by the book. As the frustrations of landowners, hunters, and politicians accumulated, a few complained to elected legislators and commissioners about the biologist's recommendations and *modus operandi*.

Many of the commissioners understood and appreciated the old political adage "crap rolls downhill." When they got chewed on by the governor or legislators or constituents, they in turn chewed on the TGD's director. The director lectured the division chief, who cautioned the field biologists. The field biologists looked around for someone farther down the chain of command to dump on, and there was no one there. The buck—the crap—had come to rest at the bottom of the hill. We joked that, in compensation for having the political buck continuously placed in our laps, it should be considered a privilege that simply came along with the "big bucks" we received in salary and expenses.

When my first boss, Jim Teer, left the TGD in 1959 to return to the University of Wisconsin to complete his doctorate, I became the lead biologist and project leader for game management operations in the Edwards Plateau and Central Mineral Basin. At age twenty-four and with only two years of professional experience, I was now the man at the bottom of the hill.

My first recognition of what I came to know as "political truth" became clear to me at a public hearing in a bellwether county where the regulations for the upcoming hunting season—and the issue of antlerless deer permits—were laid out for the public. Controversy and politics ran hot and heavy. Political pressure by powerful players had not budged the recommendations of field biologists, who, in this case, were me and my colleagues.

Pressure was brought to bear on the top bureaucrats at TGD headquarters to step in and take over from the "young"—"naïve" was implied—field biologists. The field biologists who had not capitulated to political pressure of a few powerful landowners were not yet as politically skilled as they would become with time and experience.

So, in one case, the director of the TGD's Wildlife Biology Division announced that he personally would conduct the public meeting in Llano County, something less than a ringing endorsement of the field biologists involved (in this case, me). Every seat in the Llano County Courtroom was filled a half hour before meeting time. By the time the meeting started, the jury box was filled, extra chairs had been placed in the aisles, and folks were standing around the edges of the room. The fire marshal was getting nervous.

The atmosphere was tense as the director of the Wildlife Biology Division, Eugene A. Walker, called the meeting to order. He introduced himself, although he already knew many in the audience from his time doing research in the Hill Country fifteen years in the past. He announced that, given the tense state of affairs, he would conduct the meeting. A buzz went through the crowd.

Then one of the county's most influential landowners stood up in the third row. He was in his mid-sixties; stood well over six feet tall, was lean and deeply tanned with a leathery face, and had a full head of snow-white hair. The crowd went quiet in deference.

He said, "Gentlemen, we all are honored that y'all would come all the way up here from Austin. But we don't want y'all to take over the meeting. Jack Ward Thomas here" (he pointed to me) "is our biologist. We know him, and he is working hard to learn about us, our problems, and our wants and needs. We came here expecting him to conduct this meeting. I figure that's what should happen." He looked around at the audience.

A few called out, "That's right!" A few applauded. A couple moaned.

The director glared at me. Then, bless his heart, his eyes softened as he smiled, stepped back and gestured for me to take over the meeting. The patriarch smiled and sat down. Some applauded, though a few were not pleased. I verged on choking up.

Then the old fellow stood up again. "Now, I don't want you to make a mistake. Some here think Jack Ward Thomas is a genuine gold-plated sonofabitch. That may be. But he's our sonofabitch!"

The old rancher sat down to applause, mixed with considerable laughter. Once seated, he crossed his arms across his chest and winked at me. I suddenly felt tears well in my eyes as I savored the sincere, though quite left-handed compliment. From that moment on, things went ever better with the program, fostered by a changing attitude and a willingness to learn and practice new ways. "New" can be a very scary thing—"new" plus "naïve" can be even worse.

I believe being referred to as *our* sonofabitch was the best—and most welcome—compliment I was to received in over fifty years of public service. In these circumstances I don't think that even my sainted momma would have objected.

PUBLIC OPINION? CALL ROPER!

The TGD's project leaders who dealt with wildlife management issues appeared annually before the politically appointed Texas Game and Fish Commission (later the Texas Park and Wildlife Commission) to make recommendations about hunting seasons and regulations in each game management district. In turn, the commissioners were empowered to adopt, reject, or modify the recommendations. In my first appearance before the commissioners as a project leader, the pressure was on. I had just celebrated my twenty-fourth birthday.

The data assessment that backed up my recommendations was, by the standards of the day, both up to speed and labor-intensive. Every calculation had been done on a hand-crank rotary calculator. Simple statistical calculations (chi-square tests, t-tests, and simple regressions) took many hours to complete, involving several repetitions to assure accuracy.

Our maps and visual aids were equally "high-tech." We pushed the furniture in the Llano office back to the walls and used the linoleum floor as a "map table." We stuck some twenty county maps together using masking tape on the back and drew the deer census lines with variously colored crayons (along with the deer count data). Different colors delineated areas with different proposed rates of antlerless deer permits. The approved posture for doing the intricate mapping was to crawl on hands and knees over the maps after removing one's field boots to minimize the chance of footprints.

Besides the county maps, we used three-by-four-foot poster-board for data presentation. The pertinent information was hand-lettered with felt-tipped markers of various colors. As markers were significantly more expensive than crayons; their purchase required clearance from the TGD's Wildlife Biology Division's bean counter.

The maps and other visual aids were, when compared with the presentations by the other project leaders in the state, outstanding, if I do say so myself. My assistants who did the crayon and marker work over my pencil entries had done a masterful job.

My use of simple biometrics was, at least to my mind, equally impressive. All the other project leaders simply made typescript handouts that displayed year-to-year comparisons of average deer numbers using simple bar graphs. On the other hand, I regaled the commissioners with estimates of deer populations bounded by statistical confidence intervals. I then described the commissioners' choices, including various levels of risk. My God, was I on a roll!

I should have known something was amiss when, in the middle of my presentation, I noticed the TGD's executive director slumping ever further down in his chair with his arms folded across his chest. He was alternately glaring at me and studying his shoeshine. The head of the Wildlife Biology Division, my immediate boss, was looking at the floor and shaking his head from side to side.

The chairman leaned forward, parked his elbows on the table, and cleared his throat to get my attention. He was well into his sixties (which seemed ancient to me at the time), tall, lean, silver-haired, and dressed in a snappy tailor-made western-cut suit. His tooled handmade boots would have likely eaten up a couple of months of my pay.

He rapped on the table with his heavy gold Texas A&M class ring, which had a significant diamond in its center. His gravelly voice was marked by a heavy West Texas accent, "Son, that's all mighty interesting; it truly is. But let's cut to the chase. Please tell us, what's the consensus of public opinion on this matter?"

Flush with the righteousness of youth, coupled with serious political inexperience and ineptitude, I saw the question as an attempt by a political appointee to elevate "politics" over "science" as a decision-making tool. My college training had focused on making scientifically sound decisions, unbiased by the politics of the moment—"science *über alles!*"

I had yet come to grips with the fact that the commissioners, quite legitimately, made their decisions based on the "science" we provided to them, plus whatever economic, social, and political factors that they considered relevant. That was, after all, their job! What factors they considered relevant and how they weighted those factors was, quite legitimately, up to them.

We biologists proposed, as was our duty. The commissioners disposed, as was their charge, and for their reasons.

Now I was scared and trying not to show it. I cleared my throat and explained. "Mr. Chairman, I have given you the best technical information at my disposal. Based upon that information, I made my best professional recommendations as to appropriate courses of action. As I have no expertise or training as a social scientist or poll taker, I have not attempted to solicit—or measure and evaluate—public opinion. If you desire information on public opinion, I suggest you contract with Mr. Roper." Roper was the first pollster of the period who conducted polls of public opinion on political matters.

Even as that grossly stupid statement popped out of my mouth, I flinched. My retort, delivered under pressure, was arrogant, ignorant, and stupid—no less than a politically stupid trifecta! The room went quiet. I had come across as a genuine smartass, and it was too late to backpedal. In poker terminology, it was "in for a dime, in for a dollar."

The chairman tapped his gavel. "We're gonna take a little recess." He looked down at me, "Son, why don't y'all join me in the back room?" His demeanor was calm and his voice controlled. But clearly his eyes were flashing "VEXED, IRRITATED, VEXED!" As I followed him into the anteroom, I computed that the odds of my emerging gainfully employed lay somewhere between slim and none. I shut the door behind me.

The chairman turned, leaned against the conference table, and looked me right in the eye. "Son, you, by God, have massive *cojones*. I like that, especially in a young man. You remind me of me when I was your age: a little too full of yourself and with way more *cojones* than brains. But, son, I can't let you get away with being a wiseass, whether you meant to be or not.

"*Your* job is to get the information *we* want and need to make *our* decisions. I am the chairman. *My* job is to consider the information you and your folks provide. Then we will consider all the other information that we consider pertinent. Then *we* will try to make the best possible decisions for the people of the state of Texas.

"If you last long enough in your new job, you will learn that people are not overwhelmingly impressed by data—certainly not data standing alone. First, they have to trust the folks who gather, analyze, interpret, and present the data. Second, prescribed processes lend order, but people don't totally trust processes either. Third, trust is earned.

"In the end, people trust people—or they don't. That's the bottom line. And a big part of people trusting people is for the folks who make decisions to demonstrate respectful consideration of their opinions, concerns, questions, needs, and wants." Then, for the first time, he smiled and winked. "I ought to just wring your neck and watch you flop around on the floor, but I can't do that to a fellow Texas Aggie."

I thanked the Lord that I was wearing my very own Texas A&M class ring. Like a good Aggie should, he had taken note.

"Now, we are going back in the hearing room and I am going to reopen the hearing. Then I am gonna ask, 'What is the consensus of public opinion on this matter?' *Comprende?*"

I nodded "yes" and resumed shallow breathing. The chairman had given me a face-saving way to back off. I understood that my tender young ass was on the line as I followed him back into the hearing room. The commissioners resumed their seats behind the dais. I resumed mine at the witness table. The chairman sat down, rapped his Aggie ring on the table, and called the meeting to order. "Mr. Thomas, as I was saying before the break, what's the consensus of public opinion relative to your recommendations?"

I walked to the podium and stood up straight. I cleared my throat and took a deep breath. I looked each commissioner in the eye, one by one. "Mr. Chairman, I apologize that I can't give a well-considered response at this time. Within two weeks, I will submit in writing my estimation of public opinion on these matters, county by county. Mr. Chairman, I was taught in school that wildlife management is as much about dealing with people as it is about wildlife—maybe more so. Today I seemed to have forgotten that lesson. It won't happen again."

The chairman smiled and nodded, sending me a private message by twisting his Aggie class ring on his finger. "We will look forward to hearing from you. In closing, we thank you for the most thorough presentation of data on deer management that we have yet seen. This meeting is closed."

He banged his gavel. My heartbeat slowed to a more normal rhythm, and my urge to puke slowly subsided. I left the room looking for some Alka-Seltzer or Tums or anything along that line.

In the United States, wildlife biologists live and work in a democracy. Public opinion always counts; pretending otherwise is folly. Long-term success requires public consensus and appropriate compromise. The role of science and scientists is to inform and enlighten those charged with making decisions, ordinarily not to make the decisions themselves.

Sometimes when I notice my class ring, I wonder if it saved my just-budding career that day. In any case, since that time I have always thought of it as my lucky charm.

MUST BE A "JACK BIRD"

When I worked for the TGD out of Sonora, Sutton County, in 1957–1959, my assistant was a wildlife technician, Calvin Van Hoozer. Calvin had been born and raised on a ranch in Kerr County, which was part of our game management district. He could usually establish instant rapport with our landowner and rancher clients.

We were something of a Mutt and Jeff duo: I was six feet two inches tall in my socks, and he stood about five feet seven inches in cowboy boots. He was a friendly sort, but he didn't talk much unless he had something significant to say. He had worked for "the outfit" for four years and had just returned to work after a two-year stint with the U.S. Navy. Clearly, I was the greenhorn.

Wildlife technician Calvin Van Hoozer, Sonora, Texas, 1958. Calvin is seated at Jack's kitchen table, which also served as their office.

On a clear October evening in 1958, I walked a two-mile-long deer census line in the last hour of daylight. I started up the truck that had been left for me near the end of the line and drove along a pasture road to where Calvin waited after he had finished walking his census line.

Croton, an annual weed that dominated the overgrazed pasture, provided ripening seeds for mourning doves, quail, and other seed-eating birds. As we drove along, we occasionally flushed Texas nighthawks from the strips of bare ground that marked the pasture road.

I identified the bird. "That's a common nighthawk."

Calvin didn't answer.

After a pause, I noted, "The scientific, Latinized name is *Chordeiles minor*. They are in the goatsucker family—*Caprimulgidae*."

Calvin raised an eyebrow. "Really?"

"Nighthawks summer from Canada to Panama, mostly in range country. Some of them winter in Argentina."

Calvin grunted.

"They make a funny *peenting* sound when they are flying and diving to catch insects. They're really common in towns and are obvious as they catch insects attracted by streetlights. Some bats do the same thing."

Calvin looked at me out of the corner of his eye, slumped further forward in the seat, and folded his arms across his chest. His body language was clear. I should have taken note but didn't.

"They have long whiskers, sort of like a cat, which gives them increased ability to snatch insects out of the air."

Calvin pulled his sweat-stained Stetson down over his eyes and slumped even further down in the seat.

"They have long, narrow, pointed wings that let them fly fast and turn on a dime to catch flying insects. So they consume a lot more energy than birds with broader and more rounded wings. They rest on bare ground or rooftops to rest between feeding flights."

Calvin's eyes following the headlights reaching down the two tracks of the pasture road.

"What's a real surprise is the bird's mouth. When their beaks are closed, the mouth appears to be very small—a very small hooked beak is set off by the long cat-like whiskers."

Calvin uncrossed and recrossed his arms and stared out the side window at the full moon rising above the mesquite trees.

"But when they open their beaks wide, you can see they have a huge gaping mouth that makes it easier to catch insects in flight. In fact, nighthawks seem to be all mouth."

Calvin mumbled something I did not understand.

"Beg pardon?"

"I said maybe they ought to call it a 'Jack Bird.'"

We finished the drive to the camp house in silence. For the rest of the time we worked together, Calvin sometimes asked detailed questions that I answered if I could. But never again would I give him detailed answers to questions he didn't ask. The lesson? Try not to be a smartass. Most folks don't like that.

THE INDIAN BABY

Our TGD game survey crews commonly camped out when we were working more than fifty miles away from our duty stations. The allowance for groceries—three dollars per day per person—when we camped out and cooked for ourselves was, at the time, quite generous, and we ate well.

Even when we worked ten- to twelve-hour days, there were still many hours to while away hanging around camp. One way that we spent our off-duty hours, depending on the location of our camp, was searching for Native American artifacts in likely locations. With guidance from more experienced associates and a lot of practice, we became ever more skilled at locating likely aboriginal campsites and activity centers.

Before the mid-nineteenth century and subsequent overgrazing by domestic livestock and fire exclusion resulted in the invasion of shrub oak, juniper, and mesquite, the landscapes of south and west-central Texas were semiarid grasslands that supported migratory herds of buffalo (American bison). Trees were largely confined to valley bottoms and stream courses, where a combination of factors attracted humans—water, trees for firewood and shelter, wildlife for food (aquatic and terrestrial), and edible plants.

When looking for artifacts, we would search along stream courses for relatively deep pools incised in the limestone bedrock. These pools held water even when the stream ran dry and therefore were likely camping spots for aboriginal groups over the centuries. An unusual buildup in the soil profile, for example, could indicate the residue from centuries of campfires and human activity—called midden mounds (locals referred to them as "Indian mounds"). A little digging sometimes revealed layers of ash and charcoal accumulated from hundreds, perhaps thousands, of years of campfires. We often found bones of fish, deer, bison, and small mammals and sometimes shells of freshwater mussels.

Sometimes, if we were lucky, we found arrowheads, spear or atlatl points, flint knives, and scrapers. More commonly, we found chips of flint and broken flint blades and projectile points marking places where someone had knapped flint tools. Some sites yielded artifacts from more recent campers, including brass from expended cartridges, fishhooks, buttons, broken glass, or tin cans.

Digging for arrowheads was a dedicated pastime for some Texas Hill Country residents. Some aficionados accumulated hundreds of flint artifacts and displayed them in framed collections hanging on the walls of their homes and offices. Over the years, archaeologists made slow inroads in persuading amateur collectors to report "hot spots" for archaeologists to explore. In the late 1950s, still mired deep in ignorance, our work crews felt no guilt in searching out such treasures as a way to pass the lonely hours away from homes and families.

Caves and rock shelters, especially, drew our attention. Exposed layers of limestone typified the Edwards Plateau, especially the more westerly portions near the Pecos River. In some places, streams or wind had, over many centuries, eroded the limestone to form caves or shelters that had served as campsites for aboriginal peoples.

These shelter sites were commonly bone dry and had remained so for century after century. Sometimes the residues of hundreds or maybe thousands of fires had spilled from the entrances, and smoke had stained the overhangs, providing evidence of long-term human occupancy. Over time, most such sites had been "mined" in the search for artifacts.

In late summer of 1958, I led a six-man crew laying out deer census lines in areas of Crockett County, whose western boundary was the Pecos River. In the course of scouting for locations for census lines, I was intrigued by an isolated butte with a dramatic overhanging limestone cap.

Looking up from the river, I could see no cave, but the limestone face was heavily smoke-stained. It seemed likely that a portion of the original overhang had broken off and might be masking a shelter site. I asked the landowner for permission to check out the site. He nodded permission and noted that it would be a long, hot, treacherous climb "just to check out a bunch of goat turds." He was shaking his head and chuckling to himself when he walked away.

On Saturday, our day off, six of us headed up to the site. On the steep climb only a creosote bush here and there provided a foothold or handhold. When we reached the limestone cap, we found that the overhang had, as we suspected, broken away and was masking the shelter site. The area under the caprock that remained intact ranged from four to six feet in height, twenty to thirty feet in depth, and was about fifty feet long. The back walls and ceiling were deeply blackened from the smoke and soot of possibly centuries of campfires. We saw no evidence that artifact hunters had, as yet, mined the area.

Just a little scraping on the dirt-covered floor uncovered ash and charcoal. Next we found a few heavy animal bones, likely bison, that had been shattered (perhaps to expose the marrow). The powder-dry dust and soil that covered the floor had not been exposed to moisture.

Against the back wall at the deepest part of the recess, we saw a difference in the powdery soil and ash when we scraped away the surface litter. We dug away a foot or so of powdery ash-imbued soil and uncovered a limestone slab. Underneath the slab, there was a mat of desiccated leaves—we thought from a yucca plant—covering a bundle wrapped in a crumbling hide.

Inside was the desiccated mummy of a small child. We unwrapped just enough of the bundle to reveal the tiny mummy's face. As it was late in the day, we took our find and walked or slid back to level ground where we were camped. We said little as we drove back to camp. Though no law required it, at least none we knew of, we suspected that we should report our find to archaeologists at one of the state's universities and not disturb the site any further. We left the small mummy in the bed of our pickup when we stopped at the Pecos River to wash away the sweat and caked-on ash.

After supper, we sat around the campfire drinking boiled coffee and discussing our discovery. The upper part of the face of the small mummy was illuminated by the flickering firelight. Calvin was obviously uneasy. Finally I joshed him, "Hey, Calvin, not seeing ghosts, are you?"

The silence that ensued was pregnant. Then, very quietly, he answered, "Guys, this ain't like digging for arrowheads. This is different. And it just don't feel right. Not to me anyway."

The two wildlife biologists in the group, college graduates and "sure-enough scientists"—in our own minds, at least—disagreed. We explained that what we had found and what else might be found under that overhang could be a significant scientific find. Calvin didn't answer, but the set of his jaw indicated a lack of persuasion.

We decided to sleep on it. My eyes kept coming back to the mummy's face as long as I could see it in the dying firelight. I suspected others were doing the same, but no one broke the silence.

Sunday breakfast was a subdued affair. Finally, I broke the silence. "Calvin, why is turning the baby over to a university wrong? The kid has been dead for God knows how many hundreds of years. Who the hell is there to care at this late date?"

Calvin, squatting on his heels and holding his tin coffee cup in both hands to soak up the warmth, kept staring into the fire. Finally, he spoke quietly.

"Somebody cared once. See how he's all done up? He wasn't garbage. They were hurting when they wrapped him up like that, dug a hole, and put him in it. Then they drug that big flat rock all the way up here from down below and put it over him. Whoever did that loved him, just like we love our kids. So I guess I care, even if nobody else here does."

It was quiet around the campfire except for the scuffing of feet.

Calvin spoke again, even more quietly. "I wouldn't want my dead kid dug up and paraded around and gawked at like we've been doing—not now and not a hundred or a damned thousand years from now."

One of the biologists in the group was a lay preacher. I looked at him and asked half-jokingly, "Well, preacher man, what do you think?"

He looked up from poking in the fire with a stick and answered in a subdued voice. "I don't recall anything in the Good Book about something that's wrong becoming right with the passing of years. I thought a lot on this last night. I suspect we all did."

He looked around at our faces. "And I prayed on it. I just kept coming back to Jesus saying that we should "love our neighbors as ourselves" and "do unto others as we would have them do unto us.""

That settled the matter. We repaired, as best we could, the damage to the baby's funeral wrappings. We wrapped the bundle in a piece of blue nylon tarp and tied it up with nylon parachute cord.

The next morning, starting at daylight, Calvin and I clawed our way back up the bluff and placed the baby back in its rightful resting place. We replaced the flat limestone rock and returned things, as best we could, to the way they were when we stumbled onto the shelter. We didn't talk much, but clearly we were making our individual peace with our newly recognized kinsman.

Later in the week, we stopped by the landowner's ranch headquarters to thank him for his hospitality. As we were leaving, he asked in what seemed a slightly condescending tone, "Well, did you boys find anything interesting up in them rocks?"

I lied—and didn't feel a bit bad about it. "No, sir, it was just like you told us, nothing's up there but dust and goat turds. But we certainly do appreciate your hospitality and letting us take a look."

As I write this, I find myself hoping things have remained as we left them over a half century ago. Down deep, I know it will be only a matter of time until that baby is discovered again; maybe it has already happened. I suspect those discoverers will be a bit flummoxed by the blue nylon tarp and the nylon parachute cord that were somehow by someone at some time added to the baby's funeral vestments.

From that point forward, I became much more respectful and conscious of cultures that preceded mine. I could understand that my species, like all others, was on a long evolutionary journey that would face many forks in the trail that cannot be foreseen or even imagined.

THE BIGGEST RATTLER I EVER SAW

People will often get to talking about "big rattlers" when sitting around a campfire. This often leads to competitive storytelling about the biggest rattlesnake, and it never fails that the old-timer in the bunch will be asked, "What's the biggest rattler you ever saw?"

Because I was born and raised in Texas and spent some ten years working in the Edwards Plateau country, I have encountered more than a few really big diamondback rattlesnakes. And there is little doubt in my "rememory" as to which one was the biggest and scariest.

One September evening in 1958 or so, I was walking a deer census line in Llano County when I heard a rattler cut loose—and very close at hand. I froze in my tracks and starting looking around for the snake. Given the noise level of the rattling, I figured I was near to standing on the booger. Finally, I saw the snake, nearly twenty feet away. The size of the snake was the reason for the noise level!

He was right out in the open and coiled up. Obviously he figured that, having been caught out in the wide open, a threat was better than an attempted getaway. Rattlers, even very big ones, were no oddity in that part of Texas, and we often encountered one or two or more in walking a two-mile census line. But this was no ordinary rattler. This fellow was the biggest rattler I had ever seen. We had all heard stories of six-footers, but I had never seen one that came even close to that size. As I circled the coiled and agitated rattler, it came to me that here was a snake that really might be six feet long.

Ordinarily I just walked around the rattlers that I encountered, having exhausted the idea that killing them was in any way decreasing the population. Besides, I was more and more convinced that "live and let live" was a worthwhile philosophy, except for deer, quail, turkeys, rabbits, squirrels, doves, and other assorted legal and tasty game. In this case, however, dragging in a real honest-to-goodness six-foot-plus rattler into camp was a temptation too great to pass up.

In the process of scouting around for some rocks with which to do in the snake, I ran across a mesquite limb lying on the ground with a forked end. Heck, the story would be even better if I showed up with a *live* six-foot rattler!

So I poked at the coiled snake with my mesquite stick until he decided that trying for a getaway was a better option than getting poked to death. When the snake lined out for a nearby cactus patch, I pinned his head to the ground with my forked mesquite stick and reached down with my right hand and got a grip right behind his head—so close that he could not bite me—and picked him up.

Sure enough, he was as long as I was tall—and very heavy. The snake's head was the size of my fist, and he was as big around in the middle as my forearm, maybe even a tad bigger. He was not happy with his predicament and objected mightily by thrashing around with some vigor.

Okay, now I had the biggest damn rattler I ever saw in hand, and there was just a mile or so to carry him to the end of the census line where the crew would pick me up. The snake was heavy, though, and it was obvious that I couldn't carry him for a mile at arm's length. So I swung the rattler up, thinking I'd catch his body under my left arm to ease the burden. Somehow, though, I not only got the snake over my arm but also got him wrapped around my neck!

Now he had something to pull against to try to get his head loose, and pull he did. This was getting to be a bit exciting for both of us. He had my right hand—and his head—pulled up uncomfortably close to my face and neck. So my left hand joined my right in holding the snake in a two-handed grip.

To make matters worse, the old boy really, really smelled like a male rattler, and that odor was multiplied by the 100-degree-plus temperature and my rising sense of—well, let's say, "mounting concern." In short, it suddenly dawned on me that I had enjoyed about as much of the big rattler as I really cared about.

I struggled to come to some conclusion as to how to rid myself of my acquisition. The first idea that came to mind was to use my left hand to get my trusty razor-sharp pocketknife out of my pocket and cut off the snake's head. But it quickly became obvious that there was a small problem with that solution: how to get the knife out of my right pants pocket with my left hand. It didn't take me long to figure out that wasn't going to happen.

Plan B, which then came to mind, was to lie down and put the rattler's head on a rock and, using my left hand to hold another rock, smash his head in. But my right hand was getting so tired that I was afraid to disengage my left hand for fear the snake would pull his head loose while still wrapped around my neck. He wasn't getting any happier and certainly wasn't smelling any better as time passed.

Plan C was to choke the rattler to death. But either my hand was too tired or big rattlers choke poorly. The harder I squeezed, the harder the snake pulled, and the more arm-weary I became. I was now down to Plan D.

I came to the conclusion that the only way out of this mess was to get to the end of the census line, wait for the crew, and get them to extricate me from my predicament. So I set out for the county road a mile or so ahead. My pace was somewhere between a fast walk and a slow trot. I kept having a vision of tripping and landing flat on my face, or rather face first on the rattler's head. That didn't happen, and the rattler finally began to tire and ceased pulling so hard. That gave my aching hands and arms some relief.

I breathed more than one sigh of relief when I finally reached the road. But how was I to get across the barbed-wire fence? Fortunately, I spotted a cattle guard several hundred yards down the fence line. I made it there and crossed over.

I figured it could be as much as an hour before the crew truck arrived for relief. I couldn't decide whether to pray for someone else to come along first to help me from the embrace of the rattler (which would require some degree of explanation to some incredulous rancher, most of whom already thought TGD biologists were slightly more than a half bubble off plumb), or whether I should pray that one didn't come along.

Finally, I could see the lights of the crew truck coming, and I stepped out into the middle of the road festooned with a huge rattler wrapped around my neck. The truck stopped, and six of my buddies quickly gathered around me to gawk at the sight. Finally the crew chief asked, without cracking a smile, "Who has who here?"

Someone else declared—indelicately, I thought—"Man, you really stink!"

Once they figured out that I wasn't hurt, the laughter started and the wisecracks mounted. I was trembling and tired and had no sense of humor left. My pals finally managed to get me unraveled from the rattler and offered up a burlap sack into which I could deposit the snake.

But now a new problem arose. I had held the snake's head so long and so tightly that I couldn't seem to relax my grip. Finally, we got the snake in the bag, and after several ups and downs and "1-2-3's" I managed to release my grip and the snake went into the bag. The top was very rapidly tied shut.

Now all agreed that this snake was, far and away, the biggest rattler any of us had ever seen. The conversation turned to what we should do with "our" snake. I had somehow developed a strange feeling about the big rattler. We had been struggling for well over two hours, and he had never given up—and I felt really stupid. Suddenly there was no doubt in my mind as to what I should do with the big fellow.

I walked across the borrow ditch to the fence line, untied the top of the sack, and flung it over the fence. I figured any rattler that had lived long enough to get that big and still wanted to live that badly ought to have the chance to do so.

I was bigger and meaner than the rest of the crew and threatened each one of them with a thrashing if they ever told the story. Over the years, though, the story leaked out and someone would occasionally ask me, with a quizzical look, about the biggest rattler I ever saw. I would look them straight in the eye and ask, "How could you ever fall for a story like that?"

I was saved from a lot of ridicule by the fact that nobody ever really believed the story. After all, not even a wildlife biologist could be that stupid!

WE DON'T NEED NO STINKIN' INSTRUCTIONS

Taxonomists separate wild turkeys into several subspecies that exhibit differences in morphology, behaviors, and habitat utilization. The subspecies that occupies the Edwards Plateau region of Texas is known as the Rio Grande turkey.

They tend to concentrate in large numbers during fall and winter and then disperse widely in the spring for nesting. The birds begin their return to their traditional wintering areas, called "winter roosts," in early fall, accompanied by their young of the year.

The traditional hunting season for Rio Grande turkeys in most of Texas was coincident with deer season (mid-November through December), when turkeys were concentrated in their traditional winter roost areas. Only hunters who had hunting privileges on or near ranches that contained a traditional winter roost had a really good chance to harvest a gobbler. As a result, only a small fraction of adult male turkeys were legally taken by hunters—and no hens.

In East Texas and the rest of the eastern states, the Eastern wild turkey subspecies did not concentrate in winter roosts and was commonly subjected to two hunting seasons—in the fall and in the spring.

Thereby, more hunting opportunities were spread out over more area and demonstrably could be carried out without overharvest.

Hunting gobblers in the spring is a "whole 'nother thing" in terms of hunting techniques and an enhanced hunting experience relative to fall hunting seasons. The seduction of love-sick gobblers into shotgun range using calls that imitate hens requires mastery of a unique set of hunting techniques and skills.

In the late 1950s, I proposed a spring turkey-hunting season for the Edwards Plateau region. My recommendations, though quite logical in the biological sense, failed the test of politics. To justify a spring hunting season, a better understanding of seasonal distributions and behaviors of Rio Grande turkeys was essential. Experienced wildlife biologists know that any proposed change in hunting regulations that differs dramatically from the status quo, especially one of long standing, will be questioned and probably challenged initially. We needed research findings to back up recommendations for any such change.

So my associates and I trapped hundreds of turkeys while they were concentrated in various winter roosting areas. We marked the birds with numbered aluminum leg bands, one on each leg. Brightly colored markers, called "flashers" and affixed to the bands or tags, were coded to indicate where the birds were trapped. This research took place many years before the advent of radio telemetry techniques for birds.

By locating marked birds after they had dispersed for nesting in the spring or getting reports from ranchers, ranch hands, sportsmen, and game wardens, we could discern how far and to where the birds had dispersed. Marked gobblers killed by hunters informed us that nearly all turkeys returned to the same winter roost areas where they were previously trapped and banded.

Our primary means of trapping turkeys was a drop net. Thirty-by-thirty-foot nets, measuring ten feet high on the edges, were strung up from four corner posts. A central pole hoisted the net's center higher than the edges and corners. We scattered cracked corn on the ground under and around the net. Once the birds were feeding under the net, we jerked a trigger that released the net's attachments at the corners and the center. The heavy net dropped, catching turkeys underneath. At least that was how it was supposed to work.

Wild turkeys can be described as Nervous Nellies and are seemingly made up of nerve endings, eyeballs, and another pound of extraneous other stuff. On the other hand, wild turkeys do not seem especially bright, at least when it comes to being trapped. Although they can become trap-shy after several experiences with traps over a short period of time, they don't seem to remember much about traps from one year or one month to the next.

One day late in the winter trapping (post-hunting) season, we watched from our hiding place as several bands of turkeys, each including some birds that had been recently trapped and tagged, ate the bait up right up to the edge of the drop net but refused to go under the suspended net. After several days filled with similar frustrating experiences, it was obvious that we had to either find an alternative means of trapping or wrap up our trapping and marking operations for the year.

At the last general rendezvous of TGD wildlife biologists, one who specialized in waterfowl studies showed slides of his crew using a cannon net for trapping geese. A thirty-by-thirty-foot net was neatly folded in a stack of two-foot folds. The back edge was staked down, and the leading edge was attached via several feet of nylon rope to four projectiles weighing some five pounds each. Those projectiles were inserted into "cannons" with four-foot-long barrels. Steel stakes welded to the side of the barrels were driven into the ground and served to aim the cannons. Explosive charges, contained in 12-gauge brass shotgun shells loaded with black powder, went into the base of the cannons and were set off by electrical current. The projectiles pulled the net up and over the geese, and the dropping net pinned the birds underneath. My crew, given our better looks and superior intelligence, figured that if a bunch of waterfowl biologists could routinely trap geese, we could make this cannon net work to trap turkeys.

We drove the thirty-five miles into Sonora to gain access to a telephone—and some Tex-Mex chow and several ice-cold Lone Star beers. After some bartering, the goose guys agreed that we could borrow their cannon-net setup. We rolled out our bedrolls in the game warden's garage. My partner, Rodney G. "Rod" Marburger, had been a college football player and was six feet two inches tall and weighed about 200 pounds. At age twenty-two, he sported a prematurely slick bald head that earned him some ridicule, which he tolerated with grace. Marburger and I drove all day to the Gulf Coast to pick up the net setup. Then, to save a motel bill, we took turns driving and sleeping all the way back to our camp in eastern Sutton County. Budgets were tight—so tight that we routinely "squeezed nickels until the buffaloes farted," as Rod so delicately put it. We sometimes subsisted in the field on surplus military MREs (Meals Ready-to-Eat) provided by some buddies on a military reservation where we worked off and on.

We laid out the pieces that made up the cannon-net trap. We had a net; four cannons; four projectiles with rings to tie on the net; fifty brass 12-gauge shells loaded with black powder, with wires coming out where the primer would normally be; a gizmo with a plunger to generate the charge to detonate the shells; and *beaucoup* electrical wire in red, white, and yellow. We nodded at each other and smiled— hey, this didn't look too complicated. As skilled scientists and wildlife biologists, we decided to test the setup one step at a time.

Step one: test a cannon. We drove the metal stake attached to a single cannon's base into the ground so the barrel pointed straight up, put a charge in place, and hooked the wires onto the detonator. Fearful of damaging a projectile, we wadded up some newspapers and rammed them into the base of the cannon barrel. A quart glass Coke bottle served to simulate a projectile. We hooked up the detonator and crouched down behind a pickup. I yelled, "Fire in the hole!" and shoved down the plunger.

BOOM! Burning newspapers fluttered down from the sky into the dry grass as the Coke bottle disappeared into the stratosphere. As we hurried to stamp out the burning newspapers, we heard a strange warbling sound—the Coke bottle was reentering from near outer space. I dove under the pickup. Marburger took off running. The bottle landed some thirty yards away. I crawled out from under the truck and yelled at Marburger. "Where the hell did you think you were going?"

He scratched his bald pate and replied, "Well, I thought it might be harder for it to hit a moving target."

Step two: try firing a single heavy projectile. We figured an untethered five-pound projectile might go fifty to seventy yards downrange. We aimed the cannon down a dry laguna to facilitate finding the projectile. "Fire in the hole!" BOOM! We could see the projectile in the air as it arced downrange and disappeared over the live oak trees at the far end of the laguna. We spent the next half hour looking for the projectile—successfully, as luck would have it.

Step three: set up all four cannons, load the projectiles, hook them to the folded net, and go for the gusto. We had to wire the charges in the four cannons so that they would go off at the same time. As we remembered from high school shop class, the wiring could be done either in parallel or in series. For some reason, wiring in series seemed right to us. Wrong!

"Fire in the hole!" Instead of one simultaneous big BOOM we heard a quick series of explosions—BOOM! BOOM! BOOM! BOOM! The first projectile was pulled toward the center by the weight of the net; the second went straight, more or less; the third pulled a little left; the fourth pulled hard left. Only about a third of the net deployed, and what did deploy was in a massive tangle. It took about an hour to untangle, unravel, and regroup.

Step four: try again with the wiring configured the other way around. We discussed whether we should stake down the back of the net and decided "Naw, the net is so heavy that it won't all pay out."

"Fire in the hole!" This time we heard a simultaneous, very satisfying, great big BOOM! We watched slack-jawed as the net draped over a twenty-foot-tall live oak some forty yards downrange. It took an hour, alternating hacking chores, to chew down a very hard live oak with a dull hatchet and cut the net loose.

Step five: set up the cannon net in another area and finally catch us some turkeys. That area had been baited for several days, and turkeys were coming readily and regularly to scarf up the cracked corn. We set up the cannons to give the net plenty of arc and scattered fresh bait about ten yards in front of the folded net. And this time we staked down the back of the net.

Releasing a captured Rio Grande turkey, Sutton County, Texas, 1959.
The bird had been marked with numbered aluminum leg bands and a
color-coded marker.

The next day we were in our hiding place a half hour before daylight. At the crack of dawn, we heard turkeys, lots of turkeys, talking to one another and flying down from their roost trees. They wasted no time getting to the free lunch. In just a few minutes a couple dozen turkey hens were feeding where the net would land.

I pulled up the plunger handle on the detonator, winked at Marburger, whispered, "Fire in the hole," and shoved down the plunger. One big satisfying BOOM! ensued. The net arced gracefully into the sky, hit the end of its tethers, and then settled slowly to the ground. It was truly beautiful. Everything worked! Just one small problem: the net hung in the air so long that every single turkey, whether running or flying, got away.

Step six: refold the net, recharge the cannons, and change the angle of the cannons so the net would spend less time in the air. And it seemed reasonable to place the bait closer to the folded net so that the turkeys would have farther to go to escape.

We went back and crouched in our blind on the off chance that another bunch of turkeys would come to the bait. About midafternoon a couple of gobblers with long beards ambled in for a late lunch. When they were smack in the middle of the net, four feet from the edge and feeding with heads down, I pulled up the plunger handle and winked at my partner.

He gave me the old thumbs-up and whispered, "Fire in the hole." I shoved down the plunger. BOOM! The net deployed as expected, but in the process it neatly decapitated the two gobblers. The angle of the cannons was too low, and the gobblers were too close to the edge of the net. When they threw up their heads at the noise of the cannons, they very quickly went to turkey heaven.

Step seven: refold the net, recharge the cannons, change the angle of the cannons to one between the last two settings, and move the bait farther out from the edge of the net. By now, the sun was setting, and for safety's sake we unwired the detonator.

In an effort to lighten the mood, I quoted the Scarlet O'Hara character in the movie *Gone with the Wind*, "Tomorrow is another day!" As we sat around the campfire that night, we cogitated at length and could not conceive of another possible mistake we might make.

Marburger mused that if we could have thought of one, we would probably have tried it. I failed to see the humor.

We were in the blind the next morning well before daylight in hopes that some turkeys would give us another chance with this set-up. That didn't seem likely.

But to our surprise, the turkey-trappers' god smiled upon our endeavor. Less than an hour after daylight, three dozen hens were feeding on the bait where the net, *Inshallah*, would descend.

I pulled up the plunger handle and held up my crossed fingers. I tried not to even think the words "Fire in the hole!" I shoved down the plunger. BOOM! When the dust settled, we could make out thirty turkey hens flopping around under the net—we caught them all, every single one. And, better yet, all were in fine fettle.

Sometimes instructions and training by folks who know what they are doing can save a lot of time and heartbreak. On the other hand, going by the book can take the fun out of things. However, after this experience, I went for the training if and when any training was to be had.

We went on to use the both the cannon net and the drop nets to capture hundreds of turkeys over the next several years. And, I am proud to say, things went without a single glitch or a single additional "capture mortality." What we learned from the study presaged changes in the hunting regulations that allowed a spring turkey hunting season. That afforded many more hunters a chance to hunt turkeys and many more landowners a chance to garner additional and most welcome income.

Seven years later, working for the USFS in West Virginia, I used the knowledge and experience acquired trapping Rio Grande turkeys to capture Eastern wild turkeys. The objective was to determine how Eastern wild turkeys responded to even-aged timber management (clear-cutting). That study would lead to my entanglement in one of the great debates related to the management of the national forests—the clear-cutting vs. wildlife debates, which would bring the entire USFS to a major fork in the trail. One fork in the trail seemed inevitably to lead to another.

WHEN THE GAME GOES SOUR

During the decade that I worked for the TGD, I was from time to time called upon to ride shotgun with a game warden on patrols. In addition to providing backup, I was more likely just along to provide company. Those backup duties provided opportunities to learn the country and the people.

One night, in November of 1959, I was riding along in earnest pursuit of a warden buddy's favorite poacher. The warden had played cat and mouse with the guy, off and on, for a decade. In this case, the warden insisted that I be armed. Believing that the primary purpose of a weapon in such circumstances was to intimidate, I took along Big Dad's semi-ancient, double-barreled 12-gauge shotgun with exposed hammers. The Damascus twist-steel barrels had been sawed off to the legal minimum and the ends melded with a blowtorch.

We headed for a ranch road north of Sonora in Sutton County where, according to one of the warden's snitches, the miscreant had been jacklighting for a specific white-tailed buck with trophy-size antlers known to frequent the area. When darkness came, we were easing along a one-lane dirt road with our headlights off, guided only by a nearly full moon in a cloudless sky.

Then, as luck would have it, the warden's nemesis pulled out of a pasture road in front of us with his lights off. The warden pulled up within ten feet of Bad Boy's bumper and turned on his headlights and the flashing red light sitting on the dashboard. I imagined Bad Boy pissing his pants, as there was fresh blood on his bumper. The warden hit a short blast on his siren. Bad Boy turned off onto a pasture road to the left and stopped broadside to the main road.

The warden angled the patrol car so his headlights were on the poacher's car. Then he aimed his spotlight so as to blind Bad Boy. He told me to stay with the car and keep him covered. He stepped out with a six-cell flashlight in his left hand, his right hand on his holstered revolver, and ordered Bad Boy to keep his hands in sight and get out of the car—slowly. Once Bad Boy was out of the car, the warden told him to open the trunk.

The dome light in the patrol car was rigged so it would not come on when the doors were opened, so I was sitting in the dark. I quietly opened my door, stepped out, and stood behind the open door with my sawed-off shotgun in hand and pointed in Bad Boy's general direction. Bad Boy had no idea I was there.

When Bad Boy opened the trunk, the big buck he thought was dead showed significant signs of disagreement with that hypothesis. The warden instinctively holstered his .357 revolver and used both hands to get the trunk closed. Bad Boy stepped back to the open door of his car, reached in, and came out with a lever-action rifle.

I stepped out from behind the warden's car door, pointed my double-barreled shotgun right at Bad Boy's middle, cocked back both hammers, and yelled, "Gun! Gun! Look out!" As Bad Boy turned from the game warden toward the sound of my voice, the warden clobbered him over the head with his six-celled flashlight—batteries flew.

Bad Boy hit the ground like a half-empty bag of shelled corn and didn't move. The warden snatched up the rifle and flung it into the borrow ditch. He rolled Bad Boy onto his belly and handcuffed his hands behind his back. Bad Boy was making no noise of any kind—no cussing, no moaning, and no groaning. It occurred to me that he just might be dead; at least he was seriously unconscious.

I slumped back onto the car seat with my sawed-off shotgun pointed down between my feet. I was shaking, my heart was pounding, and I was sucking wind.

The warden put a hand on my shoulder. "Hey, Big Jack, relax. It's okay, it's okay."

After what seemed like an hour, I got the shotgun's hammers down into the half-cocked safety position and looked up at the warden. "Dammit! I wasn't scared he was going to shoot me! I thought he was going to shoot you! He couldn't even see me. I was coming down on both triggers when you busted his head. I was close to cutting the bastard right in half! And for what? A damned deer? This is all batshit crazy!"

To my great relief—and, I suspected, the warden's—Bad Boy started moaning and groaning, cussing, and squirming around in the caliche dust, indicating that, given the thickness of his skull, it was likely that he wasn't permanently damaged. I pulled out a bandanna and put pressure on his scalp wound—scalp wounds tend to bleed profusely. Then we took care of the deer.

The warden put Bad Boy in Bad Boy's car, in the passenger seat, with his hands manacled together under his left leg. Then he tied him down with the seat belt over his arms. I thought that was pretty clever. Logically enough, he didn't want Bad Boy bleeding all over his patrol car. The warden drove.

Still shaken, I followed along in the patrol car. I had nightly dreams for several weeks that involved staring down the barrels of the sawed-off shotgun at Bad Boy's big middle with my fingers tightening on the triggers. Those visions came to me in dreams for years. My God, what would my life have been like—and the warden's life, our families' lives, even Bad Boy's life and family—if I had put an ounce more pressure on those triggers?

Since then, I've known that unforeseen events can alter one's life forever—and in a split second. What had seemed at the outset to be an adult game of Robin Hood or Cops and Robbers came close, so very close, to somebody getting killed. Being a game warden is said to be among the most dangerous law enforcement jobs in the United States. I harbor no doubts about that!

"CAN I SHOOT ONE NOW?"

TGD wildlife biologists had no official responsibilities or authority for law enforcement. However, we were often drafted by our game warden colleagues for "a little assistance" when they needed back-up—or, sometimes, when they simply wanted company on what were usually long, lonesome, boring patrols. The payoff for the biologists included making the acquaintance of landowners, learning the lay of the land, and feeding voraciously off the warden's experience, knowledge, and contacts.

One late November evening in 1959, after a twelve-hour day manning a deer check station, I had just driven up to a camp house on the upper reaches of the Llano River. The local warden (referred to as "Dudley Do-Right" by the wildlife biologists, but never to his face) pulled in behind me with his lights off. He turned on his lights and hit his siren just as I was getting out of my pickup. The ensuing massive injection of adrenaline allowed me to set a new individual best in the straight-up jump while letting loose a string of loud obscenities in the process.

When Dudley quit laughing, he informed me that he needed a backup. As no other law enforcement folks were available, I was drafted. He allowed that, given the circumstances, it might be a good thing if I were armed. So I got my short-barreled shotgun out of my truck, grabbed a half-dozen shotgun shells loaded with No. 4 shot from the glove box, and got in Dudley's patrol car. I was dressed in government-issue gray coveralls that were well stained with the blood from dead deer I had examined during the last several days at the check station.

In Texas, hunting leases commonly include the use of a "hunt-ing cabin/lodge." Dudley told me we were headed for a ranch that had been leased by six hunters from Houston for the hunting season. To the rancher's dismay, the hunters had arrived accompanied by six "la-dies of the evening," and several of the hunters were a little drunk. A "civil disturbance" had ensued, involving some rather careless target practice and at least one fistfight. When the rancher had attempted to calm the party down, he got a fat lip for his trouble and had called the sheriff to deal with the situation.

The sheriff's dispatcher advised the rancher to stay in the main house until a law enforcement officer arrived on the scene. The deputy sheriff on duty was dealing with another dustup at the other end of the county. So the dispatcher called on the game warden. I had no idea why this dustup should be of any concern to the TGD—and certainly not to a wildlife biologist. But damned nearly any distrac-tion beat chowing down another can of cheap chili for supper, then plunking away at the cabin's resident mice with my Daisy Red Ryder BB gun.

When we arrived at our destination, we rolled in the last hun-dred yards with the engine and headlights off. Judging from the loud music, loud talk, and foul language coming from the cabin, a truly wildass party was in full swing.

The warden laid out the plan. He was in uniform, packing a pistol, and would deal with the situation. However, just in case things got out of hand, I was to wait on the porch with my trusty sawed-off shotgun. I was not to show myself unless he spoke the code words "I am not alone."

Now, there was a plan! I really liked the part about waiting on the porch. So I took up a position in the shadows where I could see inside and not be seen in return. Dudley opened the screen door and stepped into the light. He struck a pose—shoulders back, chest out, chin up, hand on the butt of his .357 "mangle-'em" revolver—and set out to establish control. One of the ladies, who looked remarkably like the old-time movie actress Mae West, walked up to Dudley. She smiled and grabbed him in the crotch. It was clearly not a sexual over-ture. Dudley flinched, just a little, and slapped her hand away. Mae put her thumb over the mouth of the bottle of beer in her hand, shook it vigorously, and sprayed Dudley full in the face. At first I was taken aback and then somewhat amused. But I wasn't Dudley.

Dudley had on a brand-new state-issued white Stetson hat that was painstakingly creased in his individual style. He was showing a bit of pique as his eyes cleared and he focused on the beer dripping from his hat brim.

That did it! It violated one of the codes of the West—never ever mess with a man's sombrero! Dudley announced that Mae was under arrest for assault on an officer of the law. The gentlemen hunters and their ladies of the evening rose from their seats and closed in a loose circle around Dudley.

Ever quick on the uptake, Dudley played his hole card. He qui-etly announced, "I am not alone." Several very slow seconds passed. Dudley repeated, a little louder, "I am not alone!"

Mae answered in a low husky voice, "No, honey, you sure as hell ain't alone. You all got lots of company—just look around!"

Dudley abandoned his cool and yelled, "*I am not alone!*"

At that, I put on a straight face, jerked open the screen door, and stepped in out of the darkness, letting the screen door slam shut behind me. I held my shotgun in the crook of my left arm—I had Marshal Matt Dillon from the TV series *Gunsmoke* in mind. To my chagrin, Mae was not impressed by my dramatic entrance, nor was her rather large and muscular male employer of the moment. It was show time! I upped the ante by cocking back a hammer on my shot-gun. That impressed nearly everyone, and all—save for Mae—stood still and got real quiet. I cocked back the second hammer. Mae looked Dudley right in the eye and sneered, "Honey, just what does your big dumb-looking buddy do?"

Dudley had recovered sufficiently to reassume his com-mand-and-control persona. He replied in a cold, controlled voice, "Lady, old Bubba there, he just kills on command."

The moment was too deliciously pregnant to resist. Without taking the shotgun out of the crook of my arm, I eased around to my left and checked that I wouldn't do any real damage. Then I dropped the hammer on the left barrel. The blast went through the sagging screen on the door and raised an impressive cloud of dust from the dirt road. Things went dead quiet. It was my turn to look Mae in the eye as I reloaded and asked Dudley, in a slow drawl with one eyebrow elevated, "Dudley, can I shoot one of 'em now?"

My suddenly very captive audience was unsure if I was drunk, not too bright, half a bubble off plumb, or all of the above, perhaps simultaneously. As I reloaded, the empty shotgun shells bounced on the floor. Mae's bravado drained from her face. She and her associates moved toward the back wall of the cabin. Dudley drew his revolver from its holster and held it down at his side. Just then, the rancher, with a double-barreled shotgun at port-arms position, appeared in the doorway. The party was definitely pooped!

The rancher didn't want to file charges. He just wanted his "guests" gone—after they paid their tab, of course. Now, there was a really good decision, I thought! I didn't figure that Dudley and I wanted to explain all of this in a court of law. I certainly didn't want to rationalize to the justice of the peace or my boss in the TGD why I was even there armed with a sawed-off shotgun and letting a round off through a screen door.

No arrests were made. Instead, Dudley took down the names, addresses, and phone numbers of the "gentlemen" and their "ladies of the evening." The men were informed that a phone call would be made to their wives if they made any more trouble. Clearly, there was more than one way to enforce the law.

The rancher gave them an hour to "pack up their shit and get their asses out of Dodge." At Dudley's rather pointed suggestion, they left adequate cash on the table to cover their tab. In less than an hour, nothing was left to testify to the incident but the cloud of dust settling on the ranch road—and, of course, the hole in the screen door.

Dudley, the rancher, his wife, and I never, so far as I know, told anybody about that evening. But now all, save for me, have passed on to their rewards. I have been long gone from Texas for over forty years. I don't know about Mae and her associates, but even if any of them are still alive, they are unlikely to be practicing their trade nor likely talking about that night. So it's my story to tell my way.

DUMB MISTAKES CAN LEAD TO SERIOUS CONSEQUENCES

In 1965, I recommended to the TGD commissioners that all deer hunters in Llano County be required to have every deer killed in the county checked at a check station—not just antlerless deer, as had been the case since hunting for antlerless deer commenced several years earlier. The objective was to determine, for the first time, something close to the actual number of deer harvested in the county, including buck deer.

We suspected that many more deer, especially bucks, were being killed than had been previously reported. Further, we believed that more complete data, even though only from one county, would reveal that deer hunting was a much bigger contributor to the economy of the Hill Country than anyone realized.

The Llano County Chamber of Commerce had, for several years, proclaimed Llano County to be "The Deer Capital of Texas." That claim was prominently displayed on a sign posted on every road entering and leaving Llano County. And, indeed, the Edwards Plateau and Central Mineral Basin was home to the densest population of white-tailed deer in North America. The county commissioners (unanimously) and landowners (mostly) supported our proposal, knowing that it would focus statewide or even national media attention on Llano County. Every landowner in the county was to inform all those who hunted on their properties that every deer—repeat, *ev-*

ery deer—killed had to go through the TGD's check station located just outside the town of Llano.

As the forty-five-day deer season wore on, the number of deer being checked exceeded all expectations and began to draw media attention. Daily reports appeared in newspapers and were announced on television and radio stations across the state. A very few hunters grumbled about the inconvenience, but the vast majority of hunters and landowners seemed pleased to participate in a history-making event. All in all, things went without a hitch.

Just before Christmas, late on a Sunday afternoon, I was manning the Llano County check station with another biologist and two technicians. All day hunters had streamed through with the deer that they had killed over the weekend. Several hunters were waiting patiently in line in their vehicles.

Then a big black Cadillac, driven by a tall, thin black man (a chauffeur by his dress), with a well-dressed white man (dare I say a "great white hunter"?) occupying the front passenger seat, turned off the main highway, drove directly to the front of the line of waiting trucks and cars, and slid to a halt in the gravel. Two buck deer were lashed down across the front fenders.

Before the dust settled, the obviously agitated and intoxicated "great white hunter" dismounted from the passenger's seat. He reached back inside and came out with a lever-action rifle. Rifle in hand, he walked up to one of the technicians checking deer and demanded, "Who the hell's in charge around here?"

The technician pointed me out and simultaneously alerted me by yelling, "Jack, this man is looking for you!" I turned and walked toward the man before I saw the rifle in his hand and noted his look of severe agitation. When I got within ten feet, he asked in a loud, demanding voice, "Do you know who the hell I am?" I had no clue as to who the hell he was—nor why I should give a damn. But it seemed unwise to be rude to an agitated drunk with a gun in hand. I stuck out my hand and said, "My name is Jack Thomas. What can I do for you?"

He ignored my outstretched hand. His reply was slurred but clearly hostile. "I don't give a tinker's damn who you are! I am Dr. So-

Weighing a buck for a disease study, 1965. This Texas Game and Fish Commission truck was rigged for performing necropsies of white-tailed deer.

and-So—that's who the hell I am! We got stopped at the county line by a smartass game warden who said we had to check our deer here or he would write me a ticket. I don't have time for this crap. Just who do you bureaucratic pissants think you are, anyway?"

Our relationship was going sour—fast. I struggled to be calm and polite. Polite is easy when a gun is pointed at your belly button, but calm is quite another matter. As soft words are said to "turneth away wrath," I said firmly but politely, "Sir, I will get your deer checked as soon as I finish with the two gentlemen in line in front of you."

Now he got right in my face. "No! Do it now! *Right now*, you government pissant!"

I turned around slowly, watching Dr. So-and-So out of the corner of my eye, to finish with the two hunters who were ahead of him. Then Dr. So-and-So questioned my ancestry, crudely and with gusto. I felt my face flush, and again I turned to face him.

He brought the rifle to bear on my belly button. Then he stepped forward and jabbed me with the muzzle and, just to emphasize his point, repeated the gesture with a bit more vigor. I backed up slowly and thought fast.

He punched me twice more with the gun's muzzle as I continued to back away. I was getting really mad and rather seriously frightened—a potentially nasty combination.

Just then, out of the corner of my eye, I saw that a Texas Highway Patrol car had pulled into the check station. Two uniformed officers—both fishing buddies of mine—had dismounted with their hands on their pistols and were headed our way.

Dr. So-and-So's lever-action rifle had an exposed hammer that had to be cocked back to enable the weapon to fire. The rifle was not cocked. Considering that I was significantly bigger, younger, and stronger than Dr. So-and-So—and cold sober and saturated with adrenalin to boot—I stopped backing up, stepped forward, pushed the muzzle to one side, and grabbed the rifle. I had one hand on the barrel and the other clamped over his hand on the stock, which kept him from cocking the piece. I jerked the good doctor toward me and secured the rifle between us with the muzzle pointed toward the sky.

Bystanders, recognizing what was happening, began to dodge every which way. I jerked him toward me and head-butted him flush in the nose.

Upon seeing what was going down, the two highway patrolmen came to my rescue—or maybe it was to Dr. So-and-So's rescue. By now, given half a chance, I was mad enough and scared enough to do the good doctor some serious bodily harm. When the blowhard realized that (1) two rather burly highway patrolmen flanked him, (2) they each had a .357 Smith and Wesson revolver in hand, and (3) I had his rifle so that he couldn't use it, the fight went out of him. Furthermore, he was bleeding profusely from what was likely a broken nose. Within seconds, the officers had him bent over the hood of their patrol car with his hands manacled behind his back. And then he was in the back seat of the patrol car on his way to the Llano County lockup.

The doc's chauffeur was scared, obviously distressed, and at a loss as to what to do. I told him to just stand by for a couple of hours until my shift ended and we would go down to the lockup so that he could talk with his boss. When we got to town, just after 10 P.M., the deputy sheriff told us that the county judge would deal with the situation first thing in the morning. Until then, the good doctor would remain in the lockup with the day's haul of drunks.

Dr. So-and-So's driver was a man of color. In those days, segregation in public accommodations was the order of the day in the Texas Hill Country. So I called my wife and explained the situation, and she invited him to come home with me to spend the night. That evening, during and after supper, he told me enough about Dr. So-and-So to convince me that he was a good family man and a fine surgeon—and a good boss. It seemed that he simply had had too much to drink and let his ego get the best of him.

The next morning I dropped in on the county judge and the district attorney. They were trying to figure out whether to charge Dr. So-and-So and on what offense. The list of potential charges included assault with a deadly weapon, "menacing" (whatever that was), and assault on a public official, all of which were felonies. Dr. So-and-So was potentially in some very deep legal doo-doo that could involve serious prison time. Given all the witnesses to his performance at the check station, conviction seemed a slam dunk.

The good doctor's lawyer showed up, having driven most of the night from Dallas. Now that Dr. So-and-So was cold sober and knew the score, he was polite, contrite, subdued, and looking for a deal. The county judge asked me if I wanted to press charges. I asked if it would be okay for me to visit with the doctor in his suite at Llano County's "Iron Bar Hotel."

When the deputy sheriff showed me to Dr. So-and-So's cell, the prisoner was sitting on the side of his bunk holding his head in his hands. He looked sick, scared, pale, hung over, and very contrite. Without the rifle in his hands and with tears in his eyes, he really didn't look all that dangerous.

When he started apologizing, he started to cry and crossed over the line into pitiful. If he were convicted of any of the more serious charges, which seemed inevitable, his life and the lives of his family would be severely and irreparably damaged. That did not consider his career as a physician.

I felt sorry for the man. Maybe I shouldn't have, but I did. I told him that if he promised me and the county judge to "get some help" for whatever problems he was having, I would forget the entire matter. He promised. We went upstairs and met with the judge, the doc's lawyer, and the district attorney. They concurred with our deal, and that was that. Charges were dismissed, and he and his driver were off for Dallas—but only after, I am pleased to report, he had checked the two deer at the check station.

I saw Dr. So-and-So a couple of times in succeeding years when he checked deer at the Llano County check station or dropped by my office in the courthouse just to say hello and drop off a bottle of the very finest bourbon. It is said that "all's well that ends well," and this, after all, had ended well. When I think on it, that incident reminds me how little things can explode into major incidents when ego, triggered by appropriate circumstance and coupled with a little too much alcohol, takes control.

For all the years since, when I could feel my anger rising at some perceived challenge to my own too-significant ego, the picture of Dr. So-and-So's tear-streaked face viewed through the bars of that jail cell would come to mind. It has always had an immediate calming effect as the air escaped from the balloon of inflated ego.

"GOOD-EYE" HABY

Medina County, Texas, was settled by central European immigrants, the vast majority from Germany, mostly in the last half of the nineteenth century. In the late 1950s many of the older folks still used German as their everyday language.

August Timmerman was Medina County's longtime game warden. He was a really big man, standing about six feet three inches tall and weighing at least 260 pounds. His ample belly hung over his belt. He had been born and raised in Medina County and spoke both fluent German and English, the latter with a distinct German accent.

Landowners by and large held him in high regard and routinely addressed him, or referred to him, as "Mister Timmerman"—heavy on the "Mister." His left hand had been amputated at the wrist when he was a young man as the result of a farm accident. The loss of a hand did not seem to impair his effectiveness as a game warden.

In the early 1960s, the TGD's cadre of wildlife biologists in the Edwards Plateau region was carrying out research on reproduction in white-tailed deer. Part of that study entailed collecting jawbones and reproductive tracts (ovaries and uteri) from does killed by hunters. It was possible to discern, from signs in the ovaries, whether the doe had been pregnant the year before and whether she was pregnant at the time of her demise. If a fetus was present, its sex and age, likely date of birth, and approximate date of conception could be determined. Examination of teeth revealed her age to the nearest year.

As hunters routinely field-dressed deer shortly after they were killed, the easiest way for biologists to obtain the reproductive tracts and jawbones was for the hunters to bring the reproductive tracts when they checked their deer at the check station and allow TGD employees to remove the jawbones.

Landowners issued hunters an antlerless deer permit, which had been provided to them by the TGD. When we issued doe tags to landowners, we made our pitch for cooperation in our study and gave them a cloth collection bag to contain the uterus and ovaries, along with an instruction sheet to be issued to hunters along with each permit. The instruction sheet had a diagram of the carcass of a female deer lying on her back with the abdomen laid open as it would be in field dressing. A large circle above the figure of the deer contained an enlarged view of the uterus and ovaries with an arrow pointing to where the organs were located in the body cavity. Using this sheet, the TGD employee who issued the doe permits described, in detail, the process for collecting the desired organs. Collection of the specimens usually took place without a hitch. Some landowners and hunters looked at us a bit quizzically as we made our sales pitch, but nearly all agreed to help.

Enter "Good-Eye" Haby.

Mr. Timmerman sat beside me at a long table in the Medina County Courthouse as I issued antlerless deer permits to landowners. His assistance was critical, as some of our clients spoke only German, and for many others it was their preferred language.

Mr. Haby, hat in hands, tentatively approached the table. He looked directly at August and did not acknowledge my presence. He nodded his head in an attenuated bow, clicked his heels, said, "*Guten Morgen*, Mister Timmerman," and took a seat. August introduced me and told me that Mr. Haby's nickname was "Good-Eye" and that I, like everyone else in Medina County, should call him by that name.

Good-Eye had a renegade eye that resisted his best efforts to aim it anywhere close to where his good eye pointed. I usually took Mr. Timmerman's advice, but in this case I considered it more respectful to stick with "Mr. Haby."

After I completed the necessary paperwork, I began my pitch about having hunters collect the reproductive tracts and jawbones from does killed on their property. The further I went into my sales pitch, the more puzzlement crept into Good-Eye's countenance. He glanced repeatedly at Mr. Timmerman with a pleading expression on his face.

For the third time, I started my pitch over, speaking slowly and distinctly. "Mr. Haby, when you issue a hunter a doe tag" (I held up a permit), "hand him this bag" (I held it up) "and this piece of paper"

(I picked it up from the table). "The paper describes exactly what the hunter should do." I laid the bag and diagram on the table. I pointed at the circle with the enlarged illustration of the uterus and ovaries. "Just have the hunter cut this part out and put it in the bag. Okay? Have the hunter tie the bag on the deer's leg. Okay? And we will get it at the check station. Okay?"

It was clearly not okay. Good-Eye was obviously puzzled and increasingly distressed. He was already nervous talking to an outlander—and a government official in a uniform at that. He leaned toward Mr. Timmerman and whispered to him in German. Clearly, Good-Eye thought what I had said was a little strange. Upon reflection, I could see why.

Mr. Timmerman sternly told Good-Eye, "*Sprechen Sie Englisch!*" and then "Speak English!"

I tried again, speaking slowly and distinctly to get my point across. Clearly, my change in voice and emphasis convinced Good-Eye that I was becoming increasingly irritated, which increased his nervousness. He began to squirm in his seat. I lowered my voice and adapted a gentler tone.

"Now, Mr. Haby, please hand each hunter a sack" (I held up a sack) "when you give him a permit" (I held up a permit).

Good-Eye looked at me first and then at Mr. Timmerman, nodded, and said, "*Ja, ja.*"

So far, so good. "Then give the hunter one of these diagrams. *Ja?*"

Good-Eye nodded assent and looked at Mr. Timmerman. Mr. Timmerman nodded. I pointed at the circle with the uterus and ovaries and went around the circle with my finger. "Have the hunter cut this part out and put it in the bag. *Ja?*"

Good-Eye nodded.

"Then tie the bag onto the deer's leg. We'll get it at the check station. *Ja?*"

I saw the light dawning in Good-Eye's good eye. He smiled and nodded. I smiled and nodded. He looked at Mr. Timmerman and nodded. Mr. Timmerman smiled back and nodded. Good-Eye smiled at Mr. Timmerman and announced in a confident tone, "Okay, *now* I understand." Success! Hallelujah!

Then Good-Eye continued, "But Mr. Timmerman, if we ain't got no scissors, can the hunter fold up the whole paper and put it in the bag?"

Timmerman studied the tiles in the ceiling of the courtroom. When I regained composure, I looked Good-Eye in the eye, nodded and smiled, and said, "Mr. Haby, that would be just fine—just fine." Mr. Timmerman also smiled and nodded.

Somehow every one of Mr. Haby's doe hunters came to the check station with the materials we wanted. However, every one of them had the instruction sheet neatly folded in the bag, along with the uterus and ovaries. Good-Eye was taking no chances.

MOST EMBARRASSING MOMENTS

It is not uncommon in the course of bull sessions around the campfire to get into telling stories involving embarrassing moments. When it came my turn, I usually related a couple of such tales because, to my mind, both were somewhat dramatic, funny enough, and most certainly embarrassing enough to qualify.

Near the end of my tour of duty with the TGD, tranquilizer guns, or Cap-Chur guns, which delivered darts that injected drugs into wild animals, were new technology. The Cap-Chur brand of guns used compressed air to propel a dart-syringe. The dart included a small explosive charge that was set off by contact, and the controlled explosion drove a plunger that pushed the drug through a large-gauge hypodermic needle. The needle had a barb, much like that on a fishhook, to temporarily hold the dart in the animal's hide during injection. As a bonus, the barb made it easier to retrieve the expensive darts so they could be reused.

TV shows of the period, documentaries and dramas, often featured the capture of various wild animals (usually in Africa) using dart-gun technology. Any experienced user would testify that such operations were not as simple, benign, and universally successful as portrayed on television. The heavy, fast-moving darts struck animals with enough force to penetrate the hide—or even break bones, albeit infrequently—and always produced extensive bruising. Some significant degree of tissue damage from the explosive injection of the drug was a given.

The drugs we used to capture animals were not, by any stretch of the imagination, tranquilizers. The use of the word "tranquilizer" was a euphemism to make the technique seem more benign and thus more acceptable to the public. The drug we used, nicotine salicylate, could be more accurately called an "immobilizer." It produced a temporary short circuit in the nervous system that lasted from several minutes to more than an hour. The immediate reaction of a deer to injection—onset was commonly one to five minutes—entailed muscle spasms that rendered the animal immobile, though fully conscious. Though effective, it wasn't pretty to watch. Users had to know what they were doing, and that only came with practice.

During the course of a year, my team accrued significant experience using Cap-Chur gun technology and had captured many white-tailed deer—while killing a few in the process. We euphemistically referred to such death as "capture mortality" (i.e., the injected animal died) decreased dramatically with the experience of the biologists involved.

One August day, I received a telephone call from a TGD official in the Austin headquarters. I had been requested to assist in the capture of a mature bull bison using Cap-Chur gun technology. (It was clearly intimated that the request came from the governor's office.) The bull had already thwarted several attempts at capture and had busted up some fences and damaged a couple of cowboys in the process, but none seriously. That made me feel so much better!

I protested that the assignment was not within the purview of a wildlife biologist in the employee of the TGD; such was certainly not part of my job description. My superior instructed me to consider this in the "other duties as assigned" category—which was included in my job description. I pointed out that my crew had no experience tranquilizing bison. Therefore, the chances of failure, including risks of "capture mortality," were significant. While my protests changed nothing, it became quite clear that "capture mortality" would not be considered a satisfactory outcome.

When we arrived at the ranch, our unease increased when we realized that we would have a small but politically prestigious audience. Once we got a look at the fully mature bull bison in question—and estimated his weight—it was clear that we did not have a large enough dart to deliver an adequate dose of nicotine salicylate to immobilize the bull with a single dart. The darts we had on hand could deliver only a maximum dose of 4 cc. I figured it would take at least a 12 cc dose to do the job—three darts' worth.

I would have to stick three darts, in quick succession, in an appropriate place—preferably the buttocks. Given the remote likelihood of achieving that, I pointed out that the chances were slim to none that we could immobilize the bison and get him into a truck without causing death or injury. My *de facto* bosses on the scene just laughed and waved off my concerns. I wondered if they would be as cavalier if I killed their prized bull bison. I could only hope.

Thinking what the hell, I loaded six 4 cc darts (to allow for several misses) with nicotine salicylate. To keep the drug from leaking out of the large-bore size of the hypodermic darts, I plugged the needles with petroleum jelly.

We located the bull resting in the shade of some small live oaks. I got out of the jeep and began a stalk. The bull watched me, casually chewing his cud. I maneuvered through the trees to get within twenty yards with a clear shot at his rump.

I kept a clear escape route in mind should my subject take umbrage at me sticking a dart in his ass and decide to stomp me into a blood puddle. It was just past midday, and the temperature had climbed to near 100 degrees. I had spent nearly an hour trying to get into position. The old bull, still not seriously disturbed, lurched to his feet and kept turning to face me and hiding his ass, my desired target.

I had let the muzzle of the Cap-Chur gun hang down as I eased around through the live oaks. The high temperature and the sun on the black barrel melted the petroleum jelly plugging the needle and, unknown to me, nicotine salicylate was dribbling into the barrel in front of the dart.

As Plan A was a bust, it was time for Plan B, whereby we would deftly encourage the bison out of the thicket and into a barbed-wire-fenced lane that the beast routinely took en route to a paddock for his water and hay ration. Once the bison was in the lane, we would ease in from behind in the rancher's open military-style jeep, with me in the right-hand seat. The driver would ease up behind the ambling bull. Then I would deliver the darts into his big broad ass as quickly as I could shoot and reload. The driver would keep me close enough to deliver the darts with some hope of success.

We enticed the bull into the lane, where he likely wanted to go anyway. When we came in close behind him, he began to walk faster. I told the driver to close up. The bull broke into a trot. The driver matched the bull's pace, and I leaned out and loosed the first dart. The compressed air that propelled the dart blew the nicotine that had dribbled into the barrel out the muzzle as a fine mist, some of which blew back into my wide-open eyes.

It burned—a lot. But focused on my task, I loaded and delivered another dart, and then another. Each time more nicotine mist blew back into my eyes.

All the darts hit the bull right smack in the butt and were hanging in his hide not more than six inches apart. Hot damn! But was the dosage right? Would the bull go down? Would he die? My eyes were burning.

The bull veered out of the lane into the adjacent field, taking out the fence in the process. The jeep driver followed. Just in case, I loaded another dart. My eyes were really burning now—and streaming tears. I was suddenly overtaken by waves of nausea. The bull slowed to a weaving walk and began to stagger. He stopped, wobbled, made a half circle, and went down on his brisket. The jeep slid to a stop some ten yards away.

The driver and I got out quickly and headed for the downed bison. Then my muscle control went haywire, and with my nose I plowed a groove in the soft dirt—seasoned with bison and cow shit. Though I couldn't fully control my muscles and my heart was pounding, my mind was working. I knew the nicotine mist in my eyes had been quickly absorbed. I was suffering the same effects as the bull, but to a lesser degree, I hoped!

I could hear the instructor who had brought me up to speed in the use of Cap-Chur gun and nicotine salicylate. "Gentlemen, there is no known antidote to injected nicotine salicylate. Humans are especially sensitive. In case of an accidental injection, you have only a couple of minutes to react. So stop everything you are doing, quickly drop your pants and drawers, spread your legs, put your head between your legs, and kiss your happy ass good-bye." He laughed. I laughed. We all laughed. Just now, it didn't seem so damn funny.

Fortunately, the nicotine salicylate in my eyes was only a fraction of the dose in one of the darts. My problems with muscle control passed in a few minutes; the nausea and pounding heart lasted a while longer.

The bison survived. I retrieved the darts hanging from his ass. The crew had winched him up into a truck's bed. He was propped upright to make breathing easier. He was quickly on his way to a new home on another ranch.

I was certain of one thing: for damned sure, I was out of the bison-catching business, no matter who ordered what! On the way home, we went into the first beer joint we came across and downed a couple of cheap beers—on me. I got my first toast as maybe the first wildlife biologist who "tranquilized" himself and lived to tell about it.

My second most embarrassing moment came when I was using orally administered drugs to capture deer. The Texas A&M veterinarians I worked with on wildlife disease studies had numerous contacts in the pharmaceutical development business. They had obtained an orally ingested tranquilizer for testing. The company and the veterinarians thought the drug might prove useful in the capture of wild ungulates, especially deer. They were looking for a chance to test the drug under realistic field conditions. The drug was called Tranimal (a combination of "tranquilizer" and "animal"), or simply "T," as we called it. T was advertised to render animals immobile with a very low level of "capture mortality."

We figured that maybe we could wet some shelled corn, roll it in T powder, and feed it to deer. We tried this on two penned white-tailed deer who ate doctored corn without hesitation and went down in a stupor within a quarter hour. Now, based on our broad experience, we considered ourselves ready for a field test.

I knew a rancher who had a herd of more than 100 axis deer in a 200-acre enclosure. A species of deer native to the Indian subcontinent, axis deer were privately owned "exotic" wildlife and therefore not under the state's jurisdiction.

The rancher was eager to capture as many axis deer as possible for sale to other game breeders and was willing to try a capture technique that would provide the "economy of scale" that would make the deal feasible. In addition to testing T as a capture technique, TGD researchers wanted a half-dozen axis deer for studies on dietary competition between axis deer and white-tailed deer.

And our veterinarian friends at Texas A&M also wanted some axis deer to use in studies of disease transmission between axis deer and white-tailed deer. We didn't have the money, or permission, to pay for the axis deer, but there were no standing instructions related to "swaps."

So we struck a deal. In our world at the time, it was more pragmatic to beg forgiveness than to seek permission. Permission could take forever, even when forthcoming. Forgiveness was required only when things went wrong and the wrong people found out about it.

The axis deer had been routinely fed shelled corn broadcast over bare, flat ground. So drugged animals should be easy to spot and retrieve when and if they were rendered immobile. The rancher was willing to accept the risk that some of the drugged animals might die—that is, a chance of significant "capture mortality" was acceptable.

Just before the scheduled test, a very rare and serendipitous event occurred. A half foot of snow fell in the night and stayed on the ground for two days. The axis deer, which most likely had never experienced snow, were wet and cold, confused, and even more eager than usual for their supplementary ration of shelled corn. We had the rancher withhold their rations for a full day to whet the edges on their appetites.

Now it was show time! We wetted twenty-five pounds of shelled corn, rolled it in T, scattered it over a patch of ground from which we had cleared the snow, and retreated to the cabs of our pickup trucks. Some fifty axis deer eagerly ate the doctored corn. Time passed— ten, twenty, thirty minutes. Here and there, deer began to stagger or stand straddle-legged with heads hanging down and drooling. Then here and there a deer swayed and dropped onto its belly. After forty minutes, most of the deer were down, with many sprawled on their sides. When lying on their sides and unable to move, deer are subject to bloating and can die from it. We could wait no longer.

We moved methodically from deer to deer, applying blindfolds to reduce stress and rolling the animals up onto their bellies. Twenty-five deer were placed in crates designed to keep them from rolling onto their sides. Three deer, less affected by the drug than the others, staggered downhill toward a creek, weaving, occasionally falling, and struggling back to their feet.

After we had all the drugged animals in proximity to the cleared feeding ground secured and safe from bloat, we headed toward the creek bottom in search of the three deer that had staggered away. Two were quickly located and secured, leaving one drugged deer to be dealt with.

The snow made it easy to track the deer that was still missing in action. When I spotted her, the drug was wearing off. However, she might still hide herself, lie down, roll onto her side, and succumb to bloat. As I ran downhill, I did not see the net wire from an old fence that was obscured by drifted snow. I snagged a foot, went airborne, and did a three-quarter turn in the air before landing flat on my back in the limestone rocks. The impact knocked the wind out of me—and knocked me a bit silly to boot.

Our biologist's code dictated that the welfare of the animals we dealt with came first. The crew went on and secured the last drugged deer. By then, I was sitting up, trying to regulate my breathing, stop the world from spinning, and focus both eyes in the same direction at the same time.

The veterinarian checked me out. At first, nothing seemed seriously wrong except for a showy nosebleed. However, after climbing halfway back up the slope to the trucks, I leaned over and coughed up some blood. The vet deemed it likely that I had busted a rib or two at least.

The TGD trucks and all hands were needed on site to load and care for the drugged deer. I wedged into the vet's Volkswagen beetle and took off for the hospital in Kerrville. When I arrived at the door to the emergency room, I was so stiff and sore that I couldn't extract my six-foot two-inch, 220-pound self, encased in too many heavy clothes, from the seemingly shrinking Volkswagen without more pain than I cared to tolerate.

I honked the horn to get some attention. When nobody came to my aid after a half hour or so, I decided to drive the seventy miles home to Llano. As time and miles passed, shock and my adrenaline high wore off. Increasing pain, mostly from the broken ribs, became a bigger and still growing problem.

Finally, I pulled into my driveway in Llano. The nosebleed had stained the front of my coveralls, and I suspected my face might be a bit ashen. And I couldn't get myself out of the Volkswagen bug. I pulled under the carport and honked. No response. I honked some more. Still no response. I leaned on the horn! Finally, Britt, my six-year-old son, looked out the window and finally recognized me as the driver. Responding to my vigorous beckoning, he came out to the car. I assured him I was more or less okay in spite of the dried blood on my face and coveralls.

He didn't seem so sure. I instructed him, "Go tell your momma that Daddy is in the driveway in Dr. Robinson's Volkswagen and can't get out. And then tell her that she should get dressed and follow Daddy to the hospital. Please."

In a few minutes Margaret came out wrapped in her bathrobe. In case nobody was around the emergency room in the wee hours—Llano was a very small town—I told her she should call the hospital to have someone meet me at the emergency room entrance. I drove off.

My doctor—the only doctor in town—and I got to the hospital at almost the same moment. I told him my story. He was laughing as he gave me a shot of good drugs and climbed into the passenger's seat to keep me company while they took effect. I told him my story while the happy juice worked its magic to the point that I could tolerate extraction from the Volkswagen.

After a couple of hours and some x-rays, the Doc concluded that I had broken ribs, a bruised lung, and a slight concussion. After he taped up my ribs, I rode home seated on the tailgate of a TGD pickup, much happier with the world and feeling no pain. That state of mind lasted only until the drugs began to wear off. As the doc had predicted, I was seriously "stoved up" for a month or so.

Three months later, persistent pain in my left leg led to a diagnosis of a healing cracked vertebrae and a herniated disc. I was embarrassed by the sheer stupidity of my accident and anticipated that the circumstances would, at the very least, warrant an official ass-chewing for carelessness and "failure to recognize an unsafe work environment." Then some bean counter in the TGD would surely ask why we were using an experimental drug to capture privately owned exotic deer in exchange for animals to use in a research project.

Likely several violations of regulations or standard operating procedures were buried in there somewhere. Just paying the bills through my private medical insurance and out of my own pocket seemed the smart thing to do.

Two years later, when I applied for a job with the USFS, I persuaded my doctor, a personal friend, to overlook the queries on the application form relative to "bad backs." That nagging back injury plagued me, off and on, for the next thirty-eight years until, finally, I gave up and underwent surgery to fuse the two vertebrae. Five years after that, a do-over required fusion of an additional two vertebrae—just part of the carefree and bucolic life of a wildlife biologist! The axis deer—all of them, I'm proud to say—fared just fine.

THE EAGLE TALONS DANGLE NO MORE

When a senior at Texas A&M in 1957, I was privileged to take a course in big game management from Dr. Olaf Charles "Charlie" Wallmo. Oddly, only eleven years later, we would be colleagues in the USFS's research division and involved in a building conflict over the impacts on wildlife of ongoing clear-cutting of large swaths of coniferous forests in Alaska—but that's another story.

Professor Wallmo had us review Aldo Leopold's *A Sand County Almanac*. (Wallmo was one of Leopold's last students.) That book would come to be recognized as a work of genius and would serve as the philosophical foundation for natural resources professionals, especially wildlife biologists. But when first published in 1949, it received little initial attention.

Upon graduation, budding wildlife biologists expected to be employed by a state or federal conservation-oriented agency where we would embark on our destined roles as technological elites caring for our nation's natural resources through application of the best available science and dispassionate application of our superior knowledge.

All twelve of us in the class panned *A Sand County Almanac* as the esoteric, somewhat maudlin ramblings of some dead college professor. Obviously, Professor Wallmo had cast pearls before swine.

Oddly, he seemed neither surprised nor disappointed. Years later, it came to me that he had simply planted a seed in hopes that it would take root and grow when our minds were adequately fertilized by ever increasing knowledge, experience, and maturity.

When I went to work for the TGD in 1957, its policy toward predators was clear and simple: kill as many as possible as quickly and efficiently as possible. The agency employed trappers for that purpose. My territory, the Edwards Plateau region, had been rendered essentially coyote-free. Wolves and bears were long gone. Cougars were extremely rare, and bobcats only a bit less so.

True or not, golden eagles were considered to be significant predators of young domestic sheep and goats, mule deer, white-tailed deer, antelope, and wild turkeys. Contract pilots flying small single-engine aircraft were commonly hired by ranchers, with the blessings of the TGD, to hunt eagles from the air. A "gunner" seated behind the pilot literally "flew shotgun" and dispatched eagles in one-sided aerial combat.

I had been on the job for just a few weeks when such a contract pilot flew into the Sonora airstrip for several days of eagle control efforts. For some reason, the pilot was without a gunner and therefore needed an assistant "with an iron stomach" who could handle "dog-fighting eagles." Just two months earlier, I had finished flying lessons in a Cessna 150 as part of my ROTC training at Texas A&M.

I was also pretty good with a shotgun and, given that combination of experience and skill set, considered myself qualified as a gunner. After a few minutes of instruction from the contract pilot (mostly related to not shooting the airplane), we were airborne. It had not occurred to me that the TGD—or my wife—might object to what I was doing.

We had been airborne less than twenty minutes when we spotted a golden eagle. The eagle twisted, turned, and dove to evade the pilot's attempts to put the plane in a position where I had a clear shot with little chance of blasting a strut, wing, wheel, or propeller. When air sickness intervened, I pleaded with the pilot to get me on the ground. When my feet were planted firmly on the tarmac and I finished puking, I orally submitted my formal resignation as a gunner and was subjected to considerable ribbing about my sterling performance. But I noted that no one volunteered to take my place.

Later that afternoon the game warden for Sutton County, Nolan Johnson, and I watched as the pilot and a new gunner with a cast-iron stomach shot an eagle out of the sky. The shotgun blast broke one of the eagle's wings, and it wheeled down and down several hundred feet with one wing extended to break the speed of the fall. We quickly found the eagle dragging a broken wing and hopping around among scattered mesquite and juniper trees.

The eagle faced us with its neck feathers elevated and the intact wing extended. It opened its beak wide and hissed—in fear or defiance or both. Its yellow and black eyes were wide open and blazing. The warden ended its misery with a shot from his .357 caliber service revolver. We tossed the carcass in the bed of my TGD pickup and drove back to the airstrip.

At the end of the day, there were three eagles down. We drove into town for a celebratory supper. Later a couple of dozen people gathered around my TGD pickup to examine the eagles. Several photos were taken of the game warden and me—in uniform—holding the dead birds with their wings fully extended and heads hanging down in death.

We drove to the town dump to dispose of the carcasses. The warden cut off the eagles' feet for trophies and handed me a pair, explaining how to fix the toes and talons around a tennis ball so that, when they dried, they would be fixed in the extended strike position. I followed his instructions and, when they dried, tied the feet together with a leather thong and hung them from the rearview mirror of my TGD pickup truck. It seemed the *macho* thing to do.

Later that year, the pilot encountered in the course of his duties the noted newspaper columnist Westbrook Pegler in Alpine, Texas. Pegler wrote a column that portrayed the TGD pilot as a dashing character—a hero even—who was risking his life to save livestock and wildlife from an onslaught of golden eagles. Pegler's column took on a life of its own, and the political backlash eventually brought an end to the aerial gunning of hawks and eagles. Even then, the eagle talons still dangled from my rearview mirror (owning or displaying eagle talons would not become illegal until years later).

Two years later, in 1959, I stayed overnight in a TGD camp house at the head of the South Fork of the Llano River in Edwards County. By the light of a kerosene lantern, I worked on a talk I was to deliver to the Lions Club in the nearby town of Junction the next day. I wanted to quote from Aldo Leopold's *A Sand County Almanac* and retrieved my copy from a bag of books I carried around in the gear box of my pickup. I was once again caught up in the beauty of the prose and reread the entire small book. And then, with eyes and mind opening wider, I read select parts again.

The essay "Thinking Like a Mountain" grasped my imagination and would not let go. Leopold, when a young USFS forester in New Mexico, described himself as "young and full of trigger itch" when he fired into a family of wolves and broke the back of an adult female. He described how he had walked up to her and watched "a fierce green fire dying in her eyes."

My golden eagle, struggling to stand and still ready to fight for its life, leapt from the recesses of memory. In this case, the "fierce fire" came from yellow and black eyes that were no less fierce and no less indicative of a reluctance to accept death. I blew out the lantern and stretched out on a folding cot on the screened-in porch. As I lay there, hands behind my head and listening to the night sounds, I pondered Leopold's words.

Charlie Wallmo's class from Texas A&M days seemed very far in the past, though only two years had gone by. I still thought about wildlife management primarily in technical terms. But now increasing experience was leading me to feel and think more broadly and deeply. Again and again, in my mind's eye, I saw the eagle hopping toward me with one wing dangling uselessly and the other extended—mortally wounded, terrified, and yet still defiant and reluctant to die. I was becoming less certain of the righteousness of my attitudes—and those of my society—toward predators such as eagles, coyotes, bears, foxes, bobcats, and cougars.

My eagle and Leopold's wolf were on my mind when I awoke; they stayed with me through my breakfast of *charqui*, dried peaches, and cowboy coffee. I sat on the edge of my cot, a tin cup of coffee in hand, and watched and listened as the night and its sounds gave way to those of the day. I knew what I needed to do.

At full light, I went to my truck, took a shovel from the gear box, and removed the eagle's talons hanging from the rearview mirror. I walked to the spring that was the seasonal headwaters of the south fork of the Llano River. Above the spring rose a limestone cliff covered with the "signatures" of those who had quenched their thirst and rested here over the centuries: pictographs left by the ancients, inscriptions by cavalrymen—Federal and Confederate, and initials of more recent sweethearts inside a heart, including mine and Margaret's.

I cut away the leather thong that bound the eagle's feet together and buried the talons under the overhang. I apologized for my ignorance and arrogance and expressed appreciation for new understanding. It was a significant fork in my attitude and evolving understanding.

I took up reading *A Sand County Almanac* every year on my birthday, pondered the messages, and discovered new wisdom and fresh insights. Over the years, I filled the margins of my hardbound copy with notes. The pearl Professor Wallmo laid before youthful swine belatedly came to be much appreciated, at least by one of those swine. I doubted that I was alone in that regard.

Forty-eight years later, when I served as the Boone and Crockett Club's endowed professor at the University of Montana, I gave that well-worn copy of *A Sand County Almanac*, with my notes in the margins, to a graduating doctoral student, Dr. Mark Steinbach. He was a fellow Texican—a second-generation wildlife biologist and an aficionado of the book, its author, and its wisdom and insights. I hoped that my notes might add to his understanding.

I wanted that book to be good in hands, and I felt very certain about Mark's hands. Maybe someday far away in the future, he would pass it on to another promising young wildlife biologist with his notes and insights added to mine. I liked that idea.

SIX SILVER BULLETS

In the fall of 1959, after only two years with the TGD, I was transferred from Sonora to Llano to become the project leader for the Edwards Plateau Game Management Survey. My boss and mentor, James G. Teer, had returned to the University of Wisconsin to complete his work on his doctorate. I couldn't believe that what I considered the *primo* job for a wildlife biologist in Texas was mine at the tender age of twenty-four with only two years of experience. It was to be the most significant fork in the trail in my forty-year career.

As an assistant project leader, I had been paid $285 per month. Project leaders, on the other hand, were paid $340 per month. That $55 difference was hugely significant. My family had grown to three, and Margaret, a teacher in the Sonora public schools, did not have a teaching job in Llano. Yet I lusted for that project leader's job, which I considered the plum position for a wildlife biologist in the TGD— maybe in the whole country—to do important work.

The director of the Game Management Division, Eugene A. "Gene" Walker, had a rule: three years' experience as an assistant project leader was required before he (there were no "she" wildlife biologists in those days) could become a project leader. As I had worked for only two years, I would be "acting" project leader, with hugely expanded responsibilities, a 200-mile move, and no increase in salary.

However, most generously, I was given permission to use the truck that was routinely used to transplant deer from place to place for a moving van—after hosing out the deer dung and urine. My associates volunteered their time on a weekend to help with the move. My wife and baby son would pay the price for my hard-earned good fortune.

I took the job anyway. It was, to my mind, the most important job for a TGD wildlife biologist and the one with the most potential.

I lusted for that job. While I was pleased and honored with my "promotion," my sense of injustice festered.

In late October, I was camped in a TGD camp house on the upper Llano River in Edwards County with Emmitt "Snuffy" Smith. Snuffy was the TGD's only game biologist without a smidge of college education. It was not clear to me that he had graduated from high school. We likened his official Game Biologist II title to a well-deserved battlefield commission.

He was a twenty-five-year veteran of the TGD and had spent many years as the team leader for trapping and transplanting whitetailed deer, mule deer, pronghorn antelope, and Rio Grande turkeys into vacant but suitable habitats. Snuffy was a master at public relations, which he called "just getting along." He had served in the Marine Corps in World War II and made the amphibious landings at Guadalcanal and Tarawa. With advancing age, he developed a figure about as round as tall, and his face bore a striking resemblance to a famous comedian of the time, Jimmy Durante. And Snuffy could be very hard on "newbie" wildlife biologists.

Snuffy was a superb camp cook. After we finished our supper of chicken-fried venison steak with mashed potatoes and gravy and the dishes washed and put away, I commenced bitching about "Walker's rule." I concluded that I was going to drive down to Austin and tell him exactly what I thought of his rule. And if that didn't work, I was going to appeal to TGD's executive director, Howard Dodgen. I asked Snuffy, "What do you think of them apples?"

Snuffy took a deep draw on his hand-rolled Bull Durham cigarette and let the smoke out slowly through his nose. "Have I ever told you the story about the two silver bullets?"

I was puzzled. "Two silver bullets?"

Snuffy took another draw on his cigarette, stubbed it out, and leaned back in his rocker with hands clasped behind his head. "Did you ever see the old werewolf movie starring Lon Chaney Jr.?"

I answered, "Sure. So what?"

Snuffy continued, "Well then, you remember that the only thing that could kill a werewolf was a silver bullet to the heart. To my mind, when you start a new job, you are issued, say, six silver bullets—no more and no less. Those bullets are to be held back to be used only to slay big, bad, hairy-assed werewolves."

Snuffy leaned forward. "Werewolves are the problems you encounter in any job, the really nasty problems. If you fire your last silver bullet and fail to kill the werewolf, your ass is grass. Right?"

I nodded, though increasingly puzzled.

"When you use a silver bullet, you want to make sure that you have a real, honest-to-god, hairy-assed werewolf square in your sights. So rule one is don't use silver bullets on pissants—*nada*, zero, none! Right?"

I nodded. Where was this headed?

"So, now, ask yourself, is Walker's rule worth a silver bullet? Is it a big, bad, hairy-assed werewolf we're talking about here? Or is it just another bureaucratic chicken-shit rule? I've worked with Walker for years—he hired me. Believe me, that man has never changed his mind in his whole life. He considers such as a sign of weakness. And he has his reasons—in his mind, good reasons—for everything he does. Hell, maybe it's the law?"

I got the point. And when I thought on it, I could tell a pissant from a werewolf.

Then old Snuffy said, "Always save the last silver bullet for yourself. Once you've fired off, say, five silver bullets without success, it's time to use the last one on yourself. Suck it up and move on." Snuffy routinely dished out good advice to promising young biologists. Off and on over the next ten years, he rode my ass, teaching me much in the process. I figured out later that Mr. Walker probably made that an unwritten part of Snuffy's job description.

The week before I left Texas to join the USFS in West Virginia, I was in the Llano River camp house where I had first met Snuffy. He had left a note for me that read, "Partner, you are going to be alright. You are the best of the young smart-ass biologists that I've raised yet." I felt prouder of that note—which I still have, framed and on my office wall—than any of the awards that came my way over the ensuing years. Snuff had died several months before I found his note.

Over the decades since, I've sometimes used a silver bullet, or all of those issued, with each new job. But before I pulled the trigger,

I thought of old Snuffy and made absolutely sure that I knew what I was doing and why.

With more education and experience, I came to recognize Snuffy's rule as akin to Occam's razor (i.e., the simplest plausible explanation is usually the best). Fire just one too many silver bullets, and it's likely time to seek resurrection in another job. Since that night in camp on the banks of the upper Llano River, I have shamelessly used the silver bullet analogy in advising agitated fledgling professional conservationists.

One of my last acts before leaving Texas for Morgantown, West Virginia, to take up my career with the USFS was to drive from my home in Llano to New Braunfels to say good-bye to Snuffy. He was fighting and near to losing a long ugly battle with cancer. His last words to me were, "Don't forget to use your silver bullets carefully—no shooting at pissants! And save one for yourself." He stood on his front porch, gaunt and no longer rotund, but still with a big grin, waving a farewell with his handkerchief. I would never encounter him again—except in my dreams.

Thirty-five years later, as chief of the USFS, I stood in the office of the secretary of agriculture in Washington, D.C., and used all my silver bullets—including the last one on myself. I submitted my retirement papers the next day.

As I left the secretary's office, I thought of old Snuffy. I trusted he was in a place where there was no need for silver bullets.

YOU'LL BE ON THE FRONT PAGE SOME DAY

Selling the idea of hunting of antlerless deer was a primary part of the job for wildlife biologists in Texas in the late 1950s. The state's major newspapers had outdoor writers whose regular columns were the "hook and bullet" crowd's source of straight poop on matters pertaining to hunting and fishing, including the surrounding politics.

Most of the outdoor writers I encountered were a pleasure to deal with, in spite of the fact that many of them were a bit pompous and not nearly as "skookum" about hunting and wildlife as they thought themselves to be. After a while on the job, some outdoor writers came to fancy themselves as experts on all matters related to fish and wildlife management. As part of their jobs, successful TGD biologists cultivated their attention—and even friendship when that proved possible.

Fred Maly, who wrote for the *San Antonio Light*, was "very much agin' it" when it came to doe killing. He considered himself the ultimate authority on all matters related to hunting and fishing in Texas, especially in the Texas Hill Country. He demonstrated no respect for what he called "smartass college-boy biologists" with their "newfangled theories" about wildlife, especially white-tailed deer and wild turkey management. The biologists I worked with privately referred to him as a "troglodyte," "Neanderthal," "knuckle-dragger," or, most frequently, "dumbass redneck."

I suffered his slings and arrows in silence as long as he confined himself to such relatively benign insults as "wet behind the ears," "overeducated idiot," and "biostitute." I didn't like his descriptive terms, but my colleagues and I took no action beyond mumbling under our breathes and grinding our teeth.

Then one day his column in the Sunday paper intimated that the TGD biologists (including me) in the Hill Country (my territory) were taking bribes in various forms from landowners in exchange for issuing doe tags to landowners not eligible for those permits. As the team leader for the wildlife biology program in the Hill Country, I considered that accusation as having seriously crossed the line between bearable and intolerable. That is, "them was fightin' words!"

I stewed on it for a couple of days. Then I decided to drive from Llano down to San Antonio to deal with the matter *mano a mano*. I called my boss, Al Jackson (who lived in San Antonio), to let him in on my intentions and to bum a place to mooch some food and lodging and a drink or two. I got to Al's place just in time for supper. I was halfway through dessert when Al asked, "What did old Fred write that got your feathers up?"

I took the clipping from my shirt pocket and handed it over. "Didn't you read this?"

"No," Al replied. "Why would I read anything that fool writes?" Now, there was a most perceptive question, but my elevated blood pressure was impairing my hearing.

Al perched his reading glasses on the end of his nose and silently read Maly's column a couple of times. He allowed as how maybe "old Fred" had gone just a tad too far this time. But then he asked, "Do you really think punching out that scrawny little pip-squeak is a good idea? You'll lose your job. Thought about that?"

"Damned straight. But enough is enough. Our bosses just let this kind of crap go, over and over. None of the county commissioners in the Hill Country will speak up for us. One of them in Kerr County—old "Boss" Peterson—is even egging him on. It's past time for somebody to stand up to the sorry little twerp and kick his sorry ass up around his ears!"

Al tipped his chair back. "No doubt you can do that—if he doesn't gut-shoot you with a pocket pistol. You are six inches taller, fifty pounds heavier, twenty-five years younger, and an old boxer to boot. You would feel better for a little while. Then you would lose your job, he might charge you with assault, and then he would likely sue you—and win. And worse than that, nothing will change. Let's face facts. Wildlife biologists, in general, are insignificant people engaged in an insignificant profession dealing by and large with insignificant things."

I felt the wind dump out of my sails, which began to luff. Al had a gentle smile and a twinkle in his eye, especially when counseling younger professionals. Now he used both. "Today, we wildlife biologists rarely read about ourselves and what we do read expect on the sports pages—the *sports pages*, for God's sake. Think about that. Who gives a rip? Like the baseball box scores, nobody remembers the articles on hunting and fishing the next day. Today's sports pages are tomorrow's garbage can liners." I felt increasingly foolish and rash and increasingly insignificant.

He continued, "Now listen, and listen real hard. *That won't always be the case!* Someday, much sooner than you think, we won't read about wildlife biologists—and what we do and say—solely on the sports pages. Way before your career is over, you will read about what we do and say on the front pages of the biggest newspapers in the country. You will read about our issues on the editorial pages. You will see our issues and actions on the TV news. Before long, wildlife biologists will routinely testify before committees of both state and federal legislatures and maybe even confer with presidents. They will head federal agencies such as the USFS and BLM. I think you may be among the wildlife biologists who will help make that happen and then be right in the middle of it. Sure, punching old Fred would feel really good. I'd love to do it myself. But it would be a really dumb move."

I wasn't sure Al was right, especially about the future of our profession, but his calm, cool lecture took the fire out of me. My brain slowly reengaged as my testosterone and adrenaline levels—a dangerous and potentially explosive combination—diminished. I thanked Mrs. Jackson for the meal and Al for the advice and drove back home to Llano.

Al was prescient. Over the next half century, things did change—and dramatically. Biologists and ecologists gained more and more influence. Newspaper stories concerned with wildlife and its welfare, even the welfare of ecosystems, appeared ever more frequently on the front pages and editorial pages of the nation's most influential newspapers, in news magazines, and on TV. Hearings before congressional committees, state and federal, dealt with wildlife matters. It became nearly impossible to click through the TV channels without encountering a program about wildlife, wildlife management, or hunting and fishing.

Al kept me out of trouble that day and allowed me to be part of the future he visualized for our profession. He saw the future through the dust storms of conflict and the fog of confusion.

Many old men have fitful dreams. Now that I am old and increasingly ill, more and more of my dreams are of the past. Sometimes, upon awakening, I fantasize how good it would have felt to punch Fred Maly flush in the snot-locker and then boot him square in the ass. But when I come full awake, I am grateful that Al Jackson talked me out of that heartfelt desire to fight back by telling me that it was always bad strategy to "get in a pissing match with a skunk," especially when he works for a newspaper that "buys ink by the barrel."

IF I NEVER HAVE TO EAT ANY MORE VENISON

I worked almost ten years for the TGD—mostly enjoyable, satisfying, and important work. I learned much from colleagues and accumulating experiences, came upon many forks in the trail in the process, and emerged as a more patient and effective wildlife biologist/conservationist. Most important, I learned to recognize and capitalize on opportunities and to ignore trivial distractions. But that decade was far from lucrative in the financial sense. It took both my wife and me working at full-time jobs and taking on part-time jobs to make close to a decent living wage.

Margaret, with two degrees in music education, temporarily gave up public school teaching when our first son was born. She believed that, if at all possible, raising happy, successful kids was facilitated by a stay-at-home mom, especially since I was routinely missing in action well more than half the time. So we cinched our belts a little tighter and made do on my rather meager salary plus what Margaret made teaching private voice and piano lessons at home and singing for various church services. In addition, I took on various odd jobs—mostly ranch work on weekends—to supplement our income.

Even then, the end of the money often arrived well before the end of the month. Then the old standbys of biscuits and gravy, beans and corn bread, rabbit, squirrel, quail, deer, wild boar, javelina, dove, and antelope sufficed. I never gave much thought to "making do" with such meals. After all, beans and corn bread or biscuits and gravy were among my favorites—and still are. We were getting by, though admittedly more than just a smidge to the skinny side.

One evening, along toward the end of the month, we were dining on red beans and corn bread with a buttermilk chaser. The six o'clock TV news was on, and a high-profile politician was holding forth on poverty in America. His punch line was, "Believe it or not, my fellow Americans, there are families who will go to bed tonight having had nothing but beans, corn bread, and buttermilk for supper."

Britt studied his plate, looked up at me quizzically, and asked, "Daddy, are we poor?"

I looked at Margaret. She looked at me with a questioning raised eyebrow. I thought a minute and cleared my throat. "Well, no, son, we're not poor. We're not rich, but we're not poor. Besides, I really like beans, corn bread, and buttermilk, don't you?"

Britt said, "No." After a decided pause, he added, "I don't much like buttermilk." The addendum made me feel better—but not much.

That night, as I lay in bed, I did some serious thinking. I had been out of college for nearly a decade. I had been fortunate enough to spend my early professional years working with, and for, one of the best wildlife biologists I would ever know—James G. "Jim" Teer. By the luck of the draw, I had been given sink-or-swim authority and heavy responsibilities much sooner than the norm. I had learned to swim—pretty damned well, I thought.

That suited me. I had always thrived on challenges and was highly motivated by both fear—or perhaps an abhorrence of failure—and addicted to the highs that came with success. By necessity, I had matured quickly as a professional and, based on the record, developed into an increasingly competent wildlife biologist and an administrator more rapidly than most. I had already passed up several opportunities to move up in the TGD's bureaucratic hierarchy and had, reluctantly but willingly, paid the price in terms of income. In the process I gained the bonus of rapidly enhanced capabilities as a field biologist and game manager as well as enhanced skills in leadership and public relations.

If I remained a field biologist with the TGD, our family's financial situation was about as good as it was going to get unless I was willing to pursue higher-paying administrative positions. Our entire savings plan amounted to buying one $25 Series E U.S. savings bond a month, meaning that $18.75 was withheld from my paycheck. Too often, it was necessary, along toward the end of the month, to cash in one of those bonds. We wanted our two sons to be able to attend college a decade or so in the future. But how? We had two old beater cars (both inherited upon the death of a parent), which we struggled to keep running. Maybe it was time for a change?

Then, from out of the blue, I got a call from Dr. Charles "Charlie" Wallmo, my favorite college professor who had left teaching at Texas A&M shortly after I graduated. After a stint as a researcher with the Arizona Game Department, he became the project leader for a USFS unit in Fort Collins, Colorado. He had a job opening for a wildlife technician, not a "scientist," and wanted me to apply.

I thought it would be great to work for Charlie. And Fort Collins seemed a great place to live, one that would afford Margaret improved opportunities for teaching music. And, most significantly, it would pay significantly more money than I was making. I promised Charlie that I would follow up.

The federal pay scale for jobs comparable to the one I held with the TGD was considerably more than what I was making. In addition, there were greatly enhanced opportunities for promotions while remaining in a field job. And there were chances for additional education and training. So I set out to explore "gettin' me one of them good federal government jobs."

That process entailed submitting appropriate forms with extensive information relative to education, experience, and professional productivity for evaluation by a panel. Though I had no graduate school education—essentially a prerequisite for a research scientist position—I had zealously conducted research and published articles in peer-reviewed journals in addition to my wildlife management duties. Working for Jim Teer instilled that in me. Maybe that would improve my chances?

Federal pay grades ran from GS-2 to GS-17. In the research divisions of natural resources agencies, the GS-5 through GS-9 grades were held by technician assistants to "professionals" (GS-11 through GS-17). GS-13 and higher grades were usually supervisors and administrators. "Scientists," however, were an exception. Grades for scientists started at GS-11 and could, depending on performance, reach GS-15, GS-16, or GS-17, though GS-16 and GS-17 were an extreme rarity in real life. I concluded that I would certainly be qualified for the wildlife technician position (a GS-9 helper to a "professional grade" scientist), the job Charlie Wallmo had in mind for me.

On a whim, when I sent in my materials, I also requested evaluation for the positions of Wildlife Biologist—Management (GS-9 through GS-12) and Research Wildlife Biologist (GS-11 and GS-12). I didn't really believe I had any real chance for a wildlife biologist's position, certainly not in research. The research wildlife biologists I knew in federal service all held doctorates.

I sent off the paper work to the Civil Service Commission in Washington, D.C. After several months, when I had almost forgotten about the whole thing, a package arrived via registered mail with my ratings. I prepared myself for a humbling experience as I tore open the package. Wrong! I had been rated, with the highest possible scores, for all of the three positions: technician, management biologist, and scientist. I immediately asked to be taken off the roster for wildlife technician positions.

Upon request from a federal agency, the Civil Service Commission creates a roster of the top three candidates for an advertised position. The agency can then chose among those candidates. I could refuse three job offers before my name was dropped from consideration: "three strikes and out." My first offer was for a position with the Army Corps of Engineers in downtown Boston, Massachusetts. I declined: strike one. The second offer was to serve as technical editor for the research branch of the U.S. Fish and Wildlife Service in Bethesda, Maryland. I neither wanted to be nor considered myself qualified to be a professional technical editor: strike two.

Then I got a call from Dr. Ken Quigley, assistant director for the USFS's Northeast Forest Experiment Station in Upper Darby, Pennsylvania. The USFS was establishing a new Research Work Unit in conjunction with West Virginia University in Morgantown. Quigley, I thought rather coldly, informed me that West Virginia and Morgantown were not considered especially attractive duty locations. And he mentioned that the first two folks offered the job had turned him down.

Even I could take a hint. I was his third choice—and one with which Quigley was not impressed. Therefore, he said, surely I wouldn't want the job. However, he was compelled to give me a call. He wanted a yes or no answer. I asked if my relocation expenses would be reimbursed, which I knew was possible. He emphatically replied, "No."

Clearly, the USFS—at least Mr. Quigley—did not want me for the job. I could see his point. But I pushed back and asked how long I had to make a decision. He reluctantly gave me until the close of business on Friday, four days away.

I went to my "thinking place" on the banks of the Llano River. The pay for the West Virginia job was right: $10,500 a year with a chance for promotions, compared with my current salary of $7,200 (a 45 percent increase) with no chances for additional income as a field biologist. I was leery—and a little scared.

First, I would likely be the only one of the research scientists in the entire USFS without an advanced degree. Second, I had been out of the sovereign state of Texas only once—to attend the U.S. Navy Reserve boot camp in San Diego, California, in 1953. Third, I had no knowledge of the ecology of the Appalachians and knew next to nothing about forestry. Finally, at the time, the news media seemed fascinated with the poverty and pollution problems in Appalachia. So superficially West Virginia didn't appear attractive as a place to raise a family.

After our boys were asleep, I asked Margaret to go for a walk. I was nervous about asking her to leave Texas and all of our blood kin and close friends. Margaret, like me, had been out of the state only once in her life—on a school trip to California when she was a senior in high school.

I went through my sales pitch. When finished, I went silent and braced for questions. She had only one. She faced me, took my hands, and looked me straight in the eye. "Jack, if you will promise me that I will never again have to cook or eat any more venison, I will go anywhere."

I gave my word and accepted the job offer.

Margaret cooked plenty of wild game after we left Texas, especially venison from deer and elk. The important thing was that, after so many lean years, she didn't have to. And that made all the difference.

CHAPTER 3

A LONG WAY FROM TEXAS:
WEST VIRGINIA AND MAINE, 1966-1973

In the fall of 1966 our family was off to West Virginia—a place we came to love. I had taken a job as a research wildlife biologist with the USFS's Forestry Sciences Laboratory, a newly established Research Work Unit (RWU) on the campus of West Virginia University in Morgantown. We had come to another fork, a really big fork, in our trail. Margaret and I had never looked back and never regretted our choice. There were wonderful adventures and huge opportunities lying ahead out there in the fog of the future.

SOME THINGS ARE BIGGER UP NORTH

I reported for my new job in Morgantown in early December of 1966. My boss, John "Jack" Gill, a wildlife biologist with twenty years of experience, had been a senior wildlife biologist for the Maine Game Department before he came to the USFS's research division. Both John and I were well published on white-tailed deer biology but knew each other only by reputation.

Gill's carefully cultivated persona was that of a spare and taciturn New Englander with a keen, dry, subtle sense of humor. Over our first month of working together, as we sized each other up, there was a touch of rivalry manifested in showing each other just how much we knew about white-tailed deer. I told my war stories in the embellished style common to Texicans. Gill matched my tales in the exaggerated, low-keyed, parsimonious style cultivated by sons of Maine.

He told me once, "You talk so damned slow, in so much detail, that I damn near go to sleep before you get to the point, if and when there is a point."

I replied that he pronounced words that ended in the letter *a* as if they ended in the letter *r*—and vice versa—which drove me nuts. Furthermore, he was damned stingy with information, which I believed was intentional, compelling me to ask "dumb questions." If I talked too much and too slowly, he talked too little and too fast.

In mid-January, we paid a work visit to Maine so that he could bring his replacement in the Maine Game Department up to speed on an ongoing study. One key aspect of the study involved trapping and tagging white-tailed deer with the intent of determining how far, and to where, the deer dispersed in the spring.

On our drive up to Maine from West Virginia, Gill expounded upon the differences in deer behavior in Texas and Maine. In Maine several feet of snow accumulates by mid-winter, forcing deer to gather in deer yards, dense evergreen woodlands with closed, or nearly closed, tree canopies that keep much of the snowfall from reaching the ground. The interior of those stands protect deer from winds that, with temperatures well below zero, magnify the effects of low temperatures and the loss of body heat to the open sky. It very seldom snows in Central and South Texas and in the rare year when a couple of inches does fall, it stays on the ground for a very few days at most. Therefore, I had zilch experience working in ass-deep snow and had only read about "deer yards."

I was admittedly a stranger in a strange land. So I was relieved to encounter something that I recognized and knew something about—trapping with Pisgah deer traps.

While working with the Texas Game Department, I had helped trap some hundreds of white-tailed and mule deer for various reasons, such as tagging, relocation, and transplants. Most were trapped using Pisgah traps, which were constructed out of wood to trap white-tailed deer on the Pisgah National Forest in Georgia in the 1930s. The up-to-date traps we used were constructed of aluminum slats that formed an oblong box some seven feet long, four feet tall, and four feet wide. Each end was a sliding door that could be lifted up and held in place by a triggering mechanism. Deer were baited into the trap, and when they tripped a trigger, both doors fell simultaneously.

When removing deer from the traps, the "catcher" knelt at one end of the trap as the deer bounded back and forth from one end of the trap to the other. When the deer hit the door where the catcher knelt, an assistant threw up the door. The catcher jumped into the trap and the door dropped behind him. When the deer hit the other end of the trap and bounded back, the catcher grabbed it around the flank and lifted its hind legs off the ground. A hip thrust into the deer's side usually assured that the kicking hind legs struck nothing but air. Then the trap door was thrown up and the catcher "walked" the deer out of the trap on its front legs, flipped it onto its side, and held it down with his body weight. Assistants attached ear tags or radio collars and drew blood samples. Then the deer was released or loaded onto a truck for transport to another area. It took a potential catcher, if quick and strong and bright enough, only a few tries to get the drill down to a routine.

Gill and I spent the morning visiting several dozen traps that had been freshly baited and set the afternoon before. The very first trap held an adult doe. Gill and his replacement rigged a net resembling a large bag or sack over the end of the trap. When they simultaneously threw up the trap door and banged on the back of the trap, the doe bolted through the opening, became entangled in the net, and went to the ground. Gill sat astride the doe while his partner punched in the ear tags and put on the color-coded collar. Then it took what seemed like a quarter hour to disentangle the doe from the net and set her free, exhausted and scared.

The next trap held another adult doe. Gill and his partner repeated the drill without a hitch. In celebration, we broke out a thermos from a backpack and warmed our hands on the tin cups as we sipped the coffee. I told the two "Mainiacs" how we removed deer from Pisgah traps in Texas. They said they were always willing to learn new tricks and declared the next extraction of a deer from a trap was to be my show—Texican style!

The next two traps had not been tripped, but trap five contained a fully mature doe. I shed my heavy wool mackinaw, mittens, and stocking cap. I coached Gill and his partner how to throw the trap door up on my signal and drop the door behind me. I crouched in front of the trap. The doe hit the door and turned back. I yelled "Now!" The door went up. I went in. The door slammed shut behind me.

Just as I expected, the doe bounced off the other end of the trap and headed back past me. I got a flank hold and pitched my hip into her side. Hee-haw! Just like the old days, except—a very big except—the doe's hind feet didn't clear the ground. Operating in four-wheel drive, she jerked me down on my knees. As I struggled to regain my feet, she commenced to buck and slammed my head into the over-head panel several times, and we went down in a tangle. Giving up crossed my mind, but there was no way in hell I could quit with the two Mainiacs looking on. Besides, being confined in the Pisgah trap *mano a mano* with the big doe, freed of any constraint and scared, didn't seem wise.

Then the doe got a hind foot in my front pants pocket and kicked opened the leg of my coveralls from hip to mid-calf. Blood from a scalp wound dripped from the tip of my nose. I still had my flank hold and struggled to get back to my feet. I got a better grip and lifted, but still couldn't get the doe's hind legs off the ground. Her next kick caught me right square in the shin. Fortunately, as I was bigger, she pooped out fast.

By God, now I had her. I yelled, "Let us out!" Up went the door, and we exited and nose-dived into a bank of loose snow. Now the doe and I had all six of our feet on the ground and were hopping around erratically. None too soon, we went down in deep snow with me on top. She was pooped. I was pooped. But I was on top and by far the heavier of the two of us—she was pinned down.

I gasped, "Now I've got her! Put in the damn ear tags!" The two Mainiacs, laughing as vigorously as they considered social-ly correct, punched in the ear tags and attached the radio collar around the deer's neck. Gill squatted down next to my head and said, "Well, Tex, we're all done and you can let her up—whenever you're ready, of course."

I was more than ready. The big doe and I were so tangled up and so pooped that "letting her up" was more easily said than done.

Finally, after the doe and I had rested a bit, we parted company. She wobbled away, too frazzled to run. I staggered to my feet with as much dignity as I could muster, blood dripping off the end of my nose.

When we got to the next trap, there was another adult doe—maybe even bigger than her predecessor—bounding back on forth in the trap. Gill looked at me quizzically and inquired if I "hankered" to demonstrate, just one more time, how Texicans removed deer from a Pisgah trap. He suggested that I might modify my technique based on experience. I allowed that I had demonstrated as much of my skill as was appropriate for one day.

Then Gill drolly asked, "Tex, do you remember Bergman's rule?"

I lifted an eyebrow. "Bergman's rule?"

"Yeah, Bergman's rule. They did teach you about Bergman's rule in that institution of supposedly higher learning you attended in Texas?"

Then it came to me. Bergman's rule states that the farther north in the range of any mammal species, the larger the body size. There was a pretty good stretch of territory between the Texas Hill Country and northern Maine.

"Gill," I drawled (intentionally), "how big do y'all reckon that old doe to be?"

He replied, "Well, up here we don't *reckon* much, but dressed weights for adult females are about 90 to 110 pounds. I *reckon*"—he wasn't letting up on me—"that should be nigh on to twice the size of a Texas Hill Country doe of the same age. So I *reckon*, given the circumstances, you did okay." Long pause. "But maybe you should keep your mouth shut and your ears and eyes open for a while. You're a long way from Texas."

Since then, when I've been tempted to pontificate on something about which I know but little, the words "Bergman's rule" spring up from the recesses of memory, and I shut up and listen. Working with John Gill taught me a lot, especially about what he called "meticulous precision" in thinking through a problem, determining a research approach, and writing up the results.

Combining what I learned from John Gill coupled with my apprenticeship under Jim Teer in Texas provided a mid-course correction in my career. Working with Teer and Gill taught me much—and much more than just about wildlife biology. As the years passed, I did my best to pass on this knowledge to the newby biologists who came under my tutelage.

"IT WON'T BREAK YOUR JAW"

In the late 1960s the USFS was increasingly embroiled in the debate over the merits of what foresters referred to as even-age timber management, whereby stands of trees are created and maintained, through clear-cutting and selective thinning, to be of the same age and size and, usually, the same species. To achieve desired stand conditions in replacement stands, existing stands of trees of suitable size are harvested as a unit through clear-cutting.

Those stands were then "regenerated" or planted to establish a new stand of trees of the same age. The stands are periodically thinned to achieve silvicultural objectives: usually faster growth of selected trees of selected species. At the time, this "new" approach, which earlier had been vigorously castigated by most foresters and conservationists, replaced single-tree or group-selection harvesting, in which single trees or small patches of trees were harvested while maintaining stands of mixed ages and often mixed species.

This shift to clear-cutting, however much sense it made economically and in terms of high-yield commercial forestry, proved hugely controversial. Most citizens who cared about forest practices viewed the initial bald patches produced by clear-cutting as downright ugly and even carelessly destructive. Many considered clear-cutting on public lands to be a return to the "bad old days" of the late 1800s and early 1900s when the forests of the eastern United States had been cut over and, in many cases, abandoned.

The shift from single-tree or group-selection logging to patch clear-cutting as a prelude to broad-scale adaption to so-called "even-age management" of the national forests in the Northeast was based on research by USFS research silviculturist Ben Roach and associates. Their work demonstrated the efficiency of the approach, but only in terms of timber production. Knowledge was lacking as to its effects on wildlife, especially white-tailed deer, turkeys, squirrels, and ruffed

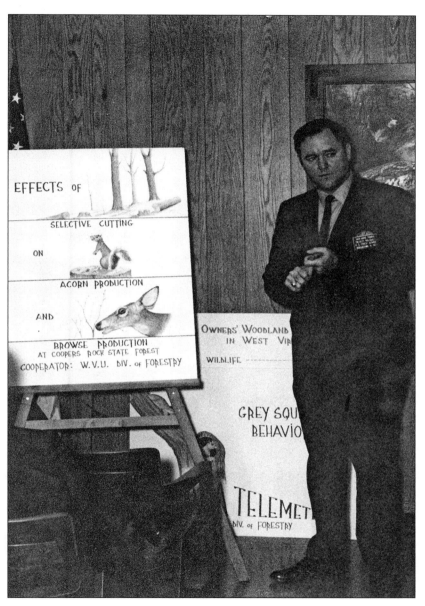

Making a presentation at the USFS Forest Sciences Laboratory,
Morgantown, West Virginia, 1968.

grouse. Little effort had been made to study the attitudes and behavior of hunters, fishers, and recreationists relative to the shift to even-age timber management. The approach was beginning to be a public relations disaster for the USFS, so far as environmentalists were concerned.

The primary assignment for the research team to which I was assigned was to fill in those gaps in knowledge. Our team was composed of project leader John D. Gill, H. Reed Sanderson, William "Bill" Healy, and Franz Pogge. R. Wayne Bailey, and James "Jimmy" Pack represented the West Virginia Conservation Commission.

Our primary study area was located on the Monongahela National Forest in the Ridge and Valley Province of southeast West Virginia. The "ridges," referred to as "mountains" by local folk, ran basically north-south and rose hundreds of feet above the intervening valleys. Such individual "mountains" were twenty to thirty miles long and two to five miles wide.

The valleys, which could support agriculture, had by and large remained in private ownership, while uplands had been purchased by the federal government, with the intent of protecting watersheds, under the auspices of the Weeks Act of 1911 and were included in the Monongahela National Forest. All of the national forests east of the Mississippi were so established.

The Weeks Act authorized the secretary of agriculture to "examine, locate, and purchase . . . forested, cutover, or denuded lands within the watersheds of navigable streams as in his judgment may be necessary to the regulation of the flow of navigable streams or for the production of timber." Land within those boundaries could be purchased from willing sellers subject to approval of the state involved. Those national forests still contained in-holdings belonging to folks whose ancestors had declined to sell their property to be included in the USFS system.

One of our study areas, Middle Mountain, was roughly encircled by a combination of state, county, USFS, and private roads. The road on the west side of Middle Mountain was a two-lane dirt road. Almost exactly halfway between the ends of the mountain was a forty-five-acre piece of private property that straddled the road with locked gates at each end. Only a few individuals had keys to the locks,

USFS personnel not being among them. Being unable to cross that property caused great inconvenience—in my opinion—for agency work crews, including ours, but the USFS had wisely made no effort to forcefully acquire a right-of-way.

All in all, I found West Virginians to be among the most polite, gentle, and genuinely nice folks that I have been privileged to know and work with. On the other hand, many "outlanders" viewed West "By God" Virginians—especially those who lived "up the hollows"—as, let's say, somewhat primitive. Some USFS folks talked in such terms about the fellow who owned and lived on the place that blocked our access. Oddly, most had never met the man.

After we had worked on Middle Mountain for several years, I gave a tour and a briefing about our research efforts to the newly assigned forest supervisor of the Monongahela National Forest. On the second day of his visit, we drove up to the locked gate. The supervisor, who was in uniform, grumbled, in somewhat colorful language, about the locked gate that inconvenienced USFS crews.

As he was talking, I reached for a ring of keys hanging from a knob on the dashboard. I singled out a key, handed it to him, and said, "I have a key."

"You have a key—and permission to use the road through this property?"

"It's okay. Just open the gate. And make sure to close it and lock it behind us." When he got back in the pickup, I cautioned him, "Let me do the talking when we get to the house."

The owner's board-and-batten, tin-roofed home was perched on the hillside above the road. I stopped in the wide place in the road at the foot of the drive up to the cabin and honked. I stepped out, retrieved a package from the gear box in the bed of the pickup, and "helloed" the cabin. (Some names used here have been changed to protect the guilty.) "Mr. Jim! Howdy! Mr. Jim! It's Jack Thomas! Can I come up!?"

An old man stepped out on the porch, snatched his cheaters off the tip of his nose, and tucked them in his shirt pocket. He waved. "Hey there, Big Jack! Come on up! Bring your friend!"

By the time we got to the porch, Mr. Jim's wife, a slightly

plump, gray-haired lady with deep blue eyes that danced with mischief, had joined him. We shook hands and exchanged pleasantries. I introduced the forest supervisor. Mr. Jim, looking my uniformed companion up and down, kindly invited us to "sit a spell."

I handed Mrs. Jim the package. "Margaret asked me to bring you this jar of preserves and to tell you she's real sorry she couldn't come along on this trip."

She smiled and nodded her pleasure. "You tell your missus that we surely thank her and hope she can come with you next time." She retreated into the cabin and came back with cups of black coffee on a wooden tray.

We sat in the rocking chairs on the porch, enjoying the shade and the slight breeze, and passed the time of day. My companion was quiet, somewhat stiff. He became a bit more relaxed only as time passed. Finally he said, "Mr. Jim, may I ask a question?"

Mr. Jim nodded.

"How is it that you gave Thomas a key to your gate—and permission to cross your property—and you won't do the same for other USFS folks?"

Mr. Jim leaned forward and indicated me with a toss of his head. "That fellow showed up at my door, after hollering first, and asked to visit. He was careful to say 'yes sir,' 'no sir,' 'please,' and 'thank you.'

"He brought me a pint of whiskey, store-bought. We passed the time of day and partook of some of that sipping whiskey. He asked me questions about the country and what I thought about this and that, just small talk.

"Then, after a bit, he told me what he was about. He asked, polite-like, if he and his boys could cross over my property from time to time to do their work. I give him his own key. Anytime they cross the property, they stop and visit for a spell. If we aren't about, they leave a note. They never put a foot on the ground between the two gates. Those boys are gentlemen. That's about all there is to it."

The supervisor asked if he could come back and visit at another time. Mr. Jim said that would be just fine. We took our leave.

Once we were in the USFS rig, the supervisor said, "Maybe I learned something today."

I looked at him out of the corner of my eye. "What's that?"

He tilted his hat back. "It won't break your jaw to say things like 'yes sir,' 'no sir,' 'please,' and 'thank you'—and mean it."

I smiled. "Your momma probably told you that. Mine surely did. You just forgot. By the way, the next time you go to see Mr. Jim, you might forget to wear your uniform and come in your civilian pickup. Uniforms and badges and green trucks with government agency decals don't impress Mr. Jim all that much. He sees himself as equal to any man, and it might help if you saw him in the same light. Folks like us are just passing through—here for a few years and then on to somewhere else. He and folks like him are part of this place. Their families have been hereabouts for way more than a century and will be here, many of them, a century from now. Likely his great-grandparents, grandparents, and parents are buried in the cemetery we passed back there. Plots are probably staked out for him and his wife. This is their place; they belong here. That's the difference—a significant difference—between us and folks like him."

STRAIN IT THROUGH YOUR TEETH!

As part of our research operations, we had laid out over fifty transects that crossed Middle Mountain from east to west. USFS and West Virginia state biologists walked these transects several times a year collecting data on vegetation and animal use.

A few folks who lived "back up the hollers" had been known, from time to time, to make a little moonshine whiskey, sometimes on national forest land. And some moonshiners had been known to display what a wildlife biologist might call "territorial defense behaviors" in regards to their stills. The West Virginia biologists had long before worked things out with the moonshiners. We used those contacts to work out our own live-and-let-live relationship.

We knew what areas to avoid. Then we made sure to let the fellow who ran the store and service station at one end of the mountain know our schedule far enough in advance so that he could get the word to the right folks. We let it be known, without really saying so, that we would have a hard time recognizing a still if and when we ever saw one.

It was, theoretically, our duty to tip off "the revenuers" if we ran into a still or had reason to know where one might be found. But it was equally clear that we were unarmed and alone when we were doing our work. Maybe once in a blue moon we encountered men in the woods with a shotgun or rifle who didn't seem to be hunting. Usually they nodded to acknowledge our presence. We nodded in return and sometimes passed the time of day.

Over the years, I made a friend of a reputed moonshiner who ran a small country store. I called him "Old Virgil" because his son, who ran the local service station, went by the name of "Young Virgil." Old Virgil, driving his beat-up pickup, sometimes stopped by at our check stations during the hunting seasons to sit by the fire, drink some of our coffee, shoot the bull, and swap lies. He told me that he figured, due to my Texican drawl, that I was most likely "a genuine good ole boy" and therefore probably worthy of acquaintance. He figured I wouldn't rat him out to the Dudley Do-Rights. It was a question in the form of a statement. When I was alone manning a check station at the side of the road, it was not uncommon for Old Virgil to wheel in for a visit. Sometimes he pulled out a quart jar, wrapped in burlap fastened with binding twine, and offered me a taste of his latest batch, or a friend's. I was quick to adapt to local customs.

One January, with two feet of snow on the flats, I was part of a four-man crew staying in a USFS cabin near the southern end of Middle Mountain. We were counting the tracks of white-tailed deer, wild turkeys, and other critters that crossed the transects.

Most of the crew used snowshoes. I never got the knack of using "bear paw" snowshoes—or any other kind of snowshoes, for that matter. Too frequently I got my snowshoes tangled up, usually when headed downhill, and took a header. Some of those spills were said by observers to be spectacular. Then I would breathe deep, get the snowshoes strapped back on my feet, and try again. It didn't take me too many years to give up on snowshoes. Those who had the knack of using snowshoes got their jobs done faster and with far less energy. Such routine difficulties assured that, with deep snow on the ground, I would be the last member of the survey team to reach the road and get picked up. My colleagues would be at the cabin toasting their feet by the fire and passing a jug by the time I came out of the woods and thumbed a ride, trying to arrive before it occurred to somebody to drive down the road to look for me.

One winter's eve, just shy of sundown, I crawled over the fence next to the road and saw Old Virgil parked in his ancient "crummy" (logger and USFS jargon for an old beat-up pickup). I was pleased to see him, more especially his old truck. Warming up and getting a ride to the cabin was on my mind. Old Virgil, lanky and with longish, graying black hair and beard to match, was leaning against the front fender. He was obviously feeling very little pain.

"Hey there, Big Jack! Get in the truck and warm your toes. By God, you sounded like a pissed-off bull moose coming down off that hill. I couldn't wait to see what was coming out of the woods!" He mimicked my mumbling and cursing as I waded through the knee-deep crusted snow. I tried to laugh through my embarrassment.

Once in the warm cab, I was simultaneously freezing (my hands and feet and face) and frying (my upper body, which was encased in a T-shirt, waffle-weave long underwear, a wool shirt, a down vest, and a wool jacket). Old Virgil suggested his remedy to all ills and problems. He reached under the seat and retrieved a fruit jar wrapped and tied in burlap. He saluted me with the uplifted jar, took a two-gulp "taste," shuddered, breathed deep, and handed me the jar.

"Now, Big Jack, this here will do you a world of good. This is real good stuff. It's been through the charcoal. So strain it through your teeth."

I was trying to discern if I was freezing or nearing heat stroke. So when I took my two-gulp taste of Old Virgil's concoction, I didn't bother to strain it through my teeth. I shuddered and caught a couple of deep breaths. This batch must have been closer to 120-proof than the 90-proof standard for most "store-boughten" liquor. I wiped the tears from my eyes and the snot dripping from my numb, cold-reddened nose. Old Virgil slapped me on the back and observed that he never did see much sense in drinking a whole lot of water to get a little bit of whiskey.

We each had another couple of "tastes," being careful to strain for bits of charcoal. Then, after an hour or so of solving the problems of the world, Old Virgil delivered me to the gate to the USFS camp house. The folks inside knew that Old Virgil had arrived by the sound an eight-cylinder engine makes when only seven cylinders, sometimes six, are firing regularly and the muffler is holey.

I walked into the cabin wearing only the top of my long johns above the waist. I dropped my hat and my clothes, down to my long johns, in the middle of the floor and headed for my bunk. I didn't want to interrupt the poker game.

One of the players asked, "Did Old Virgil give you a taste or two of the good stuff?"

"Several tastes," I replied. "It must have been the *really* good stuff as I chewed or coughed up several chunks of charcoal! I think I will skip supper, gentlemen. Good night." I bowed in taking my leave.

I awakened to the smell of boiling coffee and frying bacon. It was time to get up for another day of wading snow. Maybe, I thought, with any luck at all, Old Virgil will show up at the end of the line. When in Rome . . .

WHAT'S SCARY DEPENDS

The USFS operated a Job Corps Center adjacent to Middle Mountain. The center's purpose was to give young men and women having trouble getting started in life a chance to attain basic skills in reading and math while learning to operate and maintain heavy machinery. The job corpsmen were obligated to make some tangible contributions to USFS operations to offset some of the costs of their training, room and board, and walking-around money.

As part of our research operations on Middle Mountain, we had established transects at quarter-mile intervals running at right angles to the long axis of the mountain. Yellow paint on paint-can lids nailed to trees marked the lines that were used in annual measurements of vegetation, animal use, and other factors. Job corpsmen aided us in establishing those transects, usually by packing lids and two-foot steel pegs in a backpack that we used to mark transect lines.

Once the lines were in place, they helped conduct wildlife drive counts to determine populations of deer and turkeys. Drive counts were accomplished by stationing observers so that they could see the next observer around three sides of several square miles or more, usually a drainage. Then job corpsmen were spaced appropriately along the ridge of Middle Mountain to form a drive line. At the assigned time, they moved downhill, maintaining their spacing and blowing intermittently on police whistles to drive wildlife ahead of them. As the animals ran out of the surrounded area, they were counted by the stationed observers. In theory, this yielded a total count of the deer and turkeys in that drainage. It was on old technique that had been used extensively by the Civilian Conservation Corps during the Great Depression of 1929–1941. Though effective, the technique required significant manpower and had fallen into disuse when the program was disbanded at the beginning of World War II. The availability of job corpsmen, lots of them, allowed us to resurrect the technique.

I picked several job corpsmen as straw bosses. My first pick was a charismatic young black man who oozed what military types refer to as "command presence." He stood about six feet four inches tall and was every bit a "lean, mean, fighting machine," as he described himself. His handsome face, erect bearing, penetrating eyes, and deep bass voice commanded attention. He had a crop of kinky reddish hair and called himself "Big Red." He was a bright young man who, until enrolling at the center, had gotten off to a less than promising start. Now he was making up for lost time and gave evidence of turning his life around.

My research crew had a camp house and work center just across and down the road from the Job Corps Center. It was smallish but had a fully equipped kitchen and bathroom, two bedrooms, running water, and electricity. When at full strength, our USFS work crew was a bit cramped in the place, but we were not exactly roughing it. Sometimes we ate meals in the Job Corps Center's mess hall. The food was plain but hearty, and the price was way more than fair.

Big Red and I came to be friends. He and several other corpsmen sometimes hung out in our camp house when we were there over a weekend. Sometimes we helped them with their reading and basic math assignments. Once he applied himself and received some individual attention and encouragement, Big Red made rapid progress. He was soon promoted to "first sergeant" over our corpsmen work crew. I told him what I wanted, and he made it happen. The corpsmen responded more readily to Big Red than to me or my colleagues.

After several days of practice in walking crews into position for the wildlife drives, we figured we were ready for prime time. On our first drive, I blew the whistle and the drive line set off through the woods and down the hill. Each participant tooted his whistle from time to time and used hearing and vision to maintain the integrity of the drive line.

Then a corpsman flushed a black bear that chose to make its escape by running parallel to the drive line, which more or less dissolved as the corpsmen scattered. Big Red and I reorganized the crew and resumed the drive. As we proceeded on downhill, the corpsmen drifted together and formed clusters, which rendered the effort useless. When we got to the road, it took nearly an hour to account for all the corpsmen involved.

The day was a fiasco—an expensive fiasco at that. I gave my cadre of corpsmen a harsh lecture, ridiculing their fear and panic. Ridiculing a bunch of young men who were bent on maintaining a persona of *macho* fearlessness was inappropriate on my part and was deeply resented by the corpsmen. I should have known better.

Big Red tried to intervene, but my ass was chapped and I was having none of it. I slammed my hard hat down on the ground. "Dammit, guys, there's not a damned thing to be scared of. That bear was way more scared of you than you were scared of him. Nothing out here is going to hurt you! Damn!" I booted my hard hat over the borrow pit that bordered the road.

Then, in immediate response, Big Red bounced his hard hat off the ground, gave it a boot, and got in my face. "Dammit, Big Jack, these gawd-damn woods are your turf. You know what's going down. So you ain't scared of nothing. We don't know much about the woods. That's a fact.

"But I'll make you a deal. I'm going home next week on leave to Detroit, *my* territory! In *my* territory, I know what and what not to be scared of, and why. But you wouldn't know a damn thing about that, would you, white boy? So why don't you come home with me to my turf? You wouldn't stray two feet from my black ass!"

Damn! Double damn! He was right. He knew he was right. Everybody within earshot knew he was right, including me. It was time for me to fish or cut bait. I said, really loud so that all could hear, "Big Red, you're right. And I was wrong—absolutely wrong! I'm sorry."

Big Red continued to glare. Then, like the sun emerging from behind a big dark cloud, he threw back his head and laughed with that deep-throated, contagious laugh of his. Soon the whole crew was laughing.

We sat down in a circle and talked things over. I told them that if we flushed a black bear on the next drive or the one after that, then the cry would go up and be repeated down the line: "Bear!" The drive would stop until the bear made its getaway. Then the call would go up: "Bear's gone!" And the drive would resume. Big Red suggested that every ten minutes three toots on the whistles would alert the corpsmen to straighten their line and adjust their spacing, correcting for the tendency to bunch up. It would also provide an excuse for all hands to breathe deep and banish the always gathering specters of boogeymen, both real and imagined.

Knowing there is nothing to be afraid of—and believing it— are two very different things. Then, by golly, we finally got it right. We did it over and over. We finally abandoned the technique, as it simply required too much effort for the information gained (i.e., the cost/benefit ratio sucked). But it was a good try—live and learn.

The Reverend Martin Luther King Jr. was Big Red's hero. He could imitate Dr. King's voice and mannerisms so well that, if I shut my eyes, I couldn't tell if it was Big Red or Dr. King. One evening I was watching the evening news at the Job Corps Center with my crew. Dr. King was speaking in Memphis where sanitation workers were on strike—most were black. There were growing demonstrations in the streets. Dr. King urged nonviolence and called upon America to live up to its ideals of justice. The civil rights movement was approaching a crisis; there was tension in his voice and deep sadness in his demeanor. I wondered if his statement "I've seen the Promised Land. I may not get there with you" was intended as prophecy.

The next day, Big Red and I, along with a regular work crew of six black corpsmen, walked into the mess hall a little late for supper. It had been drizzling off and on all day up on Middle Mountain, which had hampered us in completing our work on schedule. As we shed our rain gear and began piling into the waiting pickups, we were suddenly bathed in brilliant sunlight, and the low-hanging clouds above us reflected red and orange and purple. We paused for a few moments to take it all in. Then we mounted up and headed for the mess hall.

We were joking and laughing when we entered. We stopped and looked at each other in puzzlement when it became immediately apparent that something was seriously wrong! The usual boisterous cacophony of words and laughter and the clatter of eating utensils rising from a roomful of healthy young men eating together at the end of a long hard day was missing, replaced by a heavy, sullen, ugly silence. The tension was so great that it seemed it could have been sliced with a knife.

All of the black corpsmen were bunched together. The white corpsmen, fewer in number, were clustered in another corner of the mess hall. I immediately picked up on the atmosphere in the room. Big Red was slower on the uptake. He spread his arms and announced, in his best imitation of Dr. King: "I have been to the mountaintop! I have seen the Promised Land!"

The silence deepened, if that was possible, and became more ominous. What seemed to be a sea of cold piercing eyes turned toward us. The center director approached from behind and gripped our arms—Big Red's and mine—and whispered in an emotion-choked voice, "Dr. King was shot and killed in Memphis today."

Big Red recoiled as if it were he who had been shot. I said, "My God, Big Red, I am so sorry." The Job Corp Center director suggested that it might be best if I left the mess hall. Suddenly, through no will of my own, I had become a "white man"—a very white man. I felt somehow guilty and mortally ashamed, though I had done nothing wrong and abhorred what I had heard.

Big Red, his crew, and I worked together off and on over much of the next year. Somehow things were never the same between us. I don't think any of us wanted it that way. But that was the way it was.

Ever since, when I have been privileged to see a glorious sunset, especially from high above a valley floor, I think of Big Red and wonder how things turned out for him. Did he ever think of our time together on Middle Mountain? Did our times together change him as much as they changed me?

From time to time, I have encountered those words of Dr. King's: "I have been to the mountaintop! I have seen the Promised Land!" When I do, I can't help but say to myself, "So have I, so have I." It was one more fork in the trail—certainly for me and, I hoped, for Big Red and his fellows.

ALLIGATOR MOUTHS AND HUMMINGBIRD ASSES

While in Morgantown, I took full advantage of the opportunity to pursue a master's degree in wildlife biology at West Virginia University. Attaining that degree would remedy my problem, somewhat self-imposed, of being the agency's only research scientist without a graduate degree. More importantly, it gave me it gave a chance to fill in serious gaps in my education, especially as related to biometrics, forestry, ecology, and land-use planning. So I took a couple of courses each spring semester while working more than full time and endeavoring to be a decent husband and father. Fortunately, Margaret was understanding and fully supportive.

One of the courses that I looked forward to taking was Population Dynamics of Ungulates, taught by an assistant professor with a newly minted doctorate from an Ivy League university. Some students called him, though not to his face, "Herr Professor" due to his somewhat pompous and autocratic nature. Yet he was undeniably, in the technical sense, a most excellent teacher who obviously knew his stuff. I was older than Herr Professor by probably ten years.

The classroom held a dozen tiers of desks that sloped up from the deck from which Herr Professor held forth. The front of the classroom was set up with chalkboards and a screen that could be raised or lowered. I sat in the middle of the last row.

Herr Professor's lectures were superb, especially his visual aids. His examinations and work assignments were tough but well thought out and fair. As older students often are, I was a "curve buster" and held a solid A as we entered the last week of the semester. Herr Professor had never called on me in class, nor had I ever asked a question.

Then, two lectures before the end of the semester, he presented a case-history study on the population dynamics of a deer herd in Central Texas. The basis of his lecture was a monograph published by The Wildlife Society (the professional society for wildlife biologists) and titled "Ecology and Management of White-Tailed Deer in the Central Mineral Basin of Texas."

During the lecture, I must have given off unconscious signals that I disagreed with Herr Professor's interpretation. He noticed and began to lecture more and more directly to me. Then he stopped, stepped from behind the podium, and pointed his finger at me. "You! In the back row. Do you have a problem with what I am saying?"

I tried to pass it off. Any student with anything approaching a normal IQ understands that getting in a pissing match with a professor, especially in the middle of a lecture, is never a smart move. "I'm sorry; I didn't mean to be a distraction."

"You were shaking your head." He mimicked my head-shaking in an exaggerated manner. "So speak up!"

Pushed into a corner, I cleared my throat. "I don't believe that your interpretation of the material in the publication under discussion is correct."

Herr Professor, an imposing big man, walked to the edge of the front row and stood, legs spread and hands on hips, and stared up at me. "Now, just who are you to question my interpretation?"

Time seemed to shift into that slow-motion mode that comes in moments of *extremis*.

"Well?" After another brief period of silence, which seemed much longer than the first, he said again, "*Well?*"

I sat up straight and said, "I am an author of the paper you are discussing."

Herr Professor walked back to the podium and snatched up the monograph. He looked at the cover and recognized my name. He picked up his lecture notes, turned, and left the classroom. The door slammed shut behind him.

After a few moments of stunned silence, laughter broke out across the lecture hall. I was embarrassed—and appalled. We students looked at each other, shrugged, picked up our books, and left. I felt queasy. Maybe I ought to go to Herr Professor's office and apologize? But apologize for what?

That night at home, I was stewing over what had happened and worrying about what to do. There was a knock on the front door. There stood Herr Professor's department head. We lived in the same neighborhood and were well acquainted through our mutual interest and participation in Little League baseball, and I had taken his most excellent nonparametric statistics course. After we sat down, he asked, "Just what went down between you and Herr Professor?"

I told him my version. He had talked with Herr Professor and with several graduate students in the class. Our versions matched, at least close enough for government work. He told me he didn't want me to go back for the last lecture or the final exam because it might be uncomfortable for both the professor and me. "You have an A in the class," he said, "so we can assume you would do well enough on the exam to maintain your grade."

I said, "I want it understood that, in my opinion, he is a top-notch professor. He's maybe a little bit too enamored of himself, but he'll get over that. I don't want this unfortunate incident to cast any doubts on his teaching ability or on his future. I have no complaints—absolutely none. Can't we just forget about it?"

My friend and neighbor said, "It's best to let it end here. He has great potential as a teacher and a researcher. I think he learned something from this experience, as did everyone in the class. But I don't want you to go back; it's just easier this way." I nodded that I understood.

Herr Professor went on to an outstanding career and eventually became a dean in a major university. Over the years, we became half-assed friends and agreed that we had both learned something that day, but not about population dynamics.

No challenge or disagreement—or person—should ever be arrogantly dismissed. Instead, the response should be respectful. "Why do you say that? How do you interpret the data?" Such use of the Socratic method can extract information that may improve understanding and certainly improves the atmosphere for learning. Arrogance, fear, and intimidation are poor teaching techniques.

As was often the case, Big Dad came to mind. He said to me once, when we were fishing in Village Creek using cane poles and bobbers with grasshoppers for bait, "Grandson, there are times to talk and times to listen. Remember, you're not learning a damn thing when you're talking. When pushed, think and ask questions, calmly and carefully. Then listen, really listen, to the answers. Most of all, don't let your alligator mouth overload your hummingbird ass."

Now, there was good wisdom that I never forgot.

HAVEN'T WE RUN INTO EACH OTHER BEFORE?

In late June of 2010, I learned that Senator Robert Byrd of West Virginia, one of the longest-serving senators, had passed away. He had been accorded, with full justification, the unofficial title of "King of Pork" in recognition of his persistent efforts and consistent success in bringing federal dollars home to West "By God" Virginia. He was proud of the title, and we West "By God" Virginians appreciated his talents in that regard.

The billions of dollars and the jobs that came with these projects were instrumental in elevating West Virginia out of poverty and ridicule as the "hillbilly state." Federal highways and byways for which he had obtained funding connected West Virginia internally and to the rest of the nation. Federally constructed infrastructure, including federal buildings, laboratories, training facilities, correctional institutions, and hospitals, bore his stamp. West Virginia's colleges and universities benefited mightily from his patronage.

Indirectly I was a beneficiary of Senator Byrd's acumen in putting federal dollars to work in West Virginia. As the story went, the USFS brass had decided to transfer three employees in its State and Private Forestry Division from Morgantown, West Virginia, to a new federal work center in Ohio. The USFS brass paid a courtesy call on Senator Byrd, who noticed such matters when they related to his turf, and informed him of that decision. The senator merely nodded and then gently inquired how much it would take to transfer the ten or so USFS folks now stationed in Ohio to Morgantown.

About that time, the USFS was requesting additional funds for two new RWUs, to be located in Ohio to work on forestry engineering and wildlife habitat. To the USFS's surprise, buried in the next appropriations bill was "earmarked funding" for two new RWUs for forestry engineering and wildlife habitat and a new laboratory to house them on the West Virginia University campus in Morgantown. And just by chance, there was enough room in the new laboratory building to house the USFS's State and Private Forestry folks to be transferred from Ohio!

As was common with the grand openings of federal programs and projects bearing Senator Byrd's brand, he delivered the keynote address that dedicated the new Forest Sciences Laboratory. On the day of the dedication, I was assigned to handle parking in the back parking lot of the laboratory. As I would join the festivities once the lot was filled, I was spiffily dressed in my only suit and tie and black dress shoes with slick leather soles.

Unbeknownst to me, Senator Byrd and his entourage, including two rather burly security guards, along with the top brass from West Virginia University and Morgantown's political elite, had driven into the laboratory's front parking lot and entered through the front door. "Meeting and greeting" and "pressing of the flesh" was under way in the foyer. Flashbulbs were popping, and the TV and news photographers and interviewers were busy.

Just as I directed the last visitor's car into its parking spot, I looked up to see several hundred protesters against the Vietnam War coming over the hill from the direction of the main university campus. They were carrying placards and chanting antiwar slogans, some quite demeaning and insulting to the senator. In my best imitation of Paul Revere, I ducked into the building's side entrance and ran down the hall toward the foyer to raise the alarm.

Senator Byrd's security guards heard my running footsteps, assumed that I was a bad guy, and stepped out to shield the senator. I tried to stop, but my leather-soled shoes spun out on the slick waxed tile floor. I smacked into the security guards, knocking one back into the senator. Senator Byrd was spared from landing on the seat of his pants when an aide caught him under the arms.

The USFS's chief of laboratory, Homer Parker, yelled out that I was a "friendly," which likely spared me from bodily harm. West Virginia State Troopers, who were gathered in the front parking lot, intercepted the protesters and contained them at a safe distance from the building. After a few minutes, the dedication ceremony began and went off without a hitch—if the chants of the demonstrators were disregarded. When the ceremony was over, I was formally introduced to the senator. I apologized for nearly knocking him off his pins. Ever-charming Southern gentleman that he was, he brushed the matter aside. He invited me to drop by his office if I ever came to Washington, D.C. My bosses' universal visage of apprehension morphed into smiles.

Twenty-eight years were to pass before I met up with Senator Robert Byrd again. In 1994, two months after I became chief of the USFS, I went to the Senate Office Building in Washington to pay a courtesy call on the senator. I was promptly escorted into his office—after all, the USFS plays a prominent role in West Virginia. Except for his mane of now snow-white hair, the senator looked much as he did when I had last seen him in Morgantown. On the other hand, I had gone gray and bald and had added some extra pounds.

After we shook hands and settled into comfortable chairs, he studied my face. Then he wagged his finger and smiled broadly. He said, "Chief Thomas, if I am not mistaken, we have quite literally run into each other before!" Then pointing his finger at his forehead, he added, "And I know where and when." Not much got by Senator Byrd. He was a steadfast friend of the USFS as long as he considered the USFS to be a friend of West Virginia. I tried to keep it that way.

BREAKING NEW GROUND

From 1969 to 1973, I led the nation's first research unit focused on urban forestry and wildlife, appropriately named the Urban Forestry and Wildlife Research Unit (RWU). Our team was housed in the School of Forestry at the University of Massachusetts (UMass) at Amherst. By now, I was hooked on graduate school education and in my

Taking a break at the Urban Forestry and Wildlife Research Unit,
University of Massachusetts, Amherst, 1972.

"spare time" earned a doctorate in forestry and land-use planning. Those years in Amherst were good years for both me and my family, and the team's research made a difference for our profession, people, and wildlife.

In 1969 the focus of the new USFS's RWU in Amherst was urban wildlife and forestry, whatever that might turn out to be. It was my job to figure that out. I was the new project leader, though "urban" had never been my habitat and I knew but little about forestry or New England.

Neither my supervisors nor my new colleagues had a clue about what specific research we might profitably undertake. And to make matters worse we were complete strangers. Our cadre was composed of me (wildlife biologist), Dr. Brian Payne (forester and economist), Robert "Bob" Brush (landscape architect), Richard "Dick" DeGraaf (wildlife biologist), Barry Gordon (social scientist), and Thomas Moore (social scientist). We were in immediate agreement that we didn't have much more than the title of our RWU to guide us: the Urban Forestry and Wildlife Research Unit.

When we asked our immediate boss, Dr. Elwood "Dick" Shafer, assistant director of the Northeastern Forest Experiment station, for instructions, direction, and advice, he shrugged his shoulders, smiled, and expressed confidence that we would figure it out. He said that he and the USFS were depending upon us to lead the way and, quoting the first chief of the USFS, Gifford Pinchot, to "break new ground" in the process.

In short order we were to come up with a research program that would grab people's interest and imagination and thereby justify the resources expended. Oh, and by the way, we were expected to define a brand-spanking-new field in wildlife biology and forestry in the process.

UMass was growing by leaps and bounds. In response, the town of Amherst and the surrounding farms and woodlands were home to many new housing developments. Brian Payne and I were the first two hands on deck. In the process of searching for homes for our families, we noted that there were hundreds of identical houses being built in dozens of subdivisions and isolated locations. Though the houses were exactly the same except for color, their prices differed

Jack's team at the Urban Forestry and Wildlife Research Unit, 1973.
Left to right: *Brian Payne (forest economist), Robert Brush (landscape architect), Dawn Bron (technician), Janet Burrati (secretary), and Richard DeGraaf (wildlife biologist).*

widely. Why? Was it because the identical houses occurred in different settings and were landscaped differently? Did the presence and kinds and sizes of trees and other vegetation in the landscape affect the price? How could we separate out other pertinent variables, such as distances to public schools, or to shopping centers, or to the university campus, from whatever difference the trees and shrubs might make in the selling price of a home?

We came up with the idea of photographing the same model of houses in dozens of different locations. The only differences in the photos were the color of the houses and the number, species, and sizes of the trees and shrubs present. Then we hired a panel of local real estate agents to appraise the houses using only the photographs. *Voilà!* The presence of trees and shrubs added significantly to the estimated values of the otherwise identical houses. Following publication in a technical forestry journal, the results were released to the media. Payne, the lead researcher, became a minor-league celebrity as his name appeared in many newspaper and magazine stories, and he was a frequent guest on radio and TV talk shows.

In the meantime, DeGraaf was walking through neighborhoods in the early mornings and late evenings taking notes on the occurrence of various species of wildlife, especially birds. After we were reported to the police several times, we worked with the police and with local media to publicize our activities.

We were not surprised to find that wildlife (primarily birds, but also squirrels, rabbits, and insects, including butterflies) was most plentiful in yards and neighborhoods with a combination of reasonably mature trees, shrubs, and gardens. But noticing a relationship and proving that relationship—while discerning the operative factors—was a different story.

Serendipitously, the National Wildlife Federation (NWF) had announced a new effort to enroll suburbanites in its Invite Wildlife to Your Backyard program and was looking for a way to kick off that campaign. The NWF published the widely acclaimed magazine *National Wildlife*. Thinking that maybe some old-fashioned *chutzpah* might work, I proposed to their leadership that we were just the folks they needed to partner with. Less than a week later, after several lengthy telephone conversations, I agreed to meet with the editor and staff at the local airport.

DeGraaf and I enlisted our landscape architect, Bob Brush, in our proposal. We were waiting when the NWF's chartered plane landed. Our first meeting took place in the back of that aircraft and then, as the temperature rose, on the tarmac under the shade of the wing. The NWF offered us the authorship of the kickoff piece for the campaign and promised to provide top-drawer artwork if we could deliver our story—with suggested sketched artwork depicting ideal habitat situations—in ten days. I looked at Brush, who would handle the artwork; he nodded yes. DeGraaf, the lead wildlife biologist, also nodded yes. I looked at the head man from the NWF and nodded yes.

I extended my hand. We shook. It was a done deal. We all knew the timelines were too tight and the plan was too sketchy. But we had agreed. Now all we had to do was make it work.

Working late into the night for the next ten days straight, we developed our plan of attack. I was responsible for developing the overall approach and the text. DeGraaf would prepare the tables showing species of vegetation in appropriate arrangements for various parts of the region and the wildlife it would be expected to attract. Brush would develop the roughed-out artwork that illustrated the points we wanted to drive home. The day before the deadline, I called the editor of *National Wildlife* and told him we were ready to meet with their graphics people and make final arrangements. The next day we met at the airport just after daylight.

We crawled into the back of the plane and went over the materials we had prepared. Within two hours we had a final agreement. The "finish artist" provided by the NWF was now the key to success, as he would convert Brush's sketches into the quality of artwork characteristic of *National Wildlife*. We had faith—we had to—that the artist could do the job while keeping the technical details intact. The next thing we would see would be the galley proofs for the magazine, too late for all but critical minor changes.

This deal was put together solely on my say-so. Now we had to convince our bosses to bend (actually ignore) the iron-clad rules relative to USFS publications: three peer reviews, editing by agency editors, clearance by supervisory personnel, final approval by editors, and a policy review by supervisory personnel. That process commonly took up to three months. Fortunately, our immediate boss, Dick Shafer, was something of a maverick and loved, understood, and encouraged what we were doing. All he asked was, "Jack, is this *really* good stuff? Do *you* back it all the way?"

"Yes."

"Okay," he said. "It's highly unlikely that we could strike such a deal with an outfit with the NWF's reputation and influence. Somehow you guys did it. But in the process you finagled a deal you were not authorized to make. So we have a choice: we back off and I chew your ass, or we go for it and break every rule in the book and then beg for forgiveness if things go sour. If it goes wrong, it will be the two of us on the hot seat. If it goes right, we can spread the credit around."

Our feature article in the next issue of *National Wildlife* served as the kickoff piece for what turned out to be the NWF's hugely popular Invite Wildlife to Your Backyard campaign. The article was reprinted and reprinted and reprinted over the next decade, so often that it became one of the most popular articles in the history of U.S. wildlife management up to that time.

Some years later, the article was revised to incorporate new information and make it applicable to other areas of the United States and Canada. It won the Outstanding Wildlife Publication Award from The Wildlife Society. Then, it was expanded into a bestselling book, *Gardening with Wildlife*, sponsored by the NWF. Our RWU's researchers wrote several of the chapters.

Our next success hinged on a fortuitous guess. Wildlife habitat and the public's enjoyment of wildlife in cities are most commonly associated with "open space," for obvious reasons. We looked at aerial photographs and maps of the large and mid-size cities of the northeastern United States and cataloged everything we thought might pass as "open space." Of course, parks and golf courses immediately jumped out, many of which were already being managed to accommodate wildlife (mostly birds) and birdwatchers.

The next most common category of "open space" was cemeteries. We separated cemeteries into categories by size and by the presence or absence of vegetation that might contribute to wildlife habitat. Many cemeteries were too small to have much of an impact as open space, and some larger cemeteries had burial plots so tightly packed that there was no space for vegetation. Others, however, were large enough to accommodate a variety of wildlife and landscaping with trees, shrubs, flowers, and grasses.

Some urban birdwatchers were way ahead of us and had been for perhaps a century. For example, Frederick Law Olmsted, the "father of American landscape architecture," had designed Mount Auburn Cemetery in Boston to be, in addition to a burial ground, a "pleasuring ground for the people." And it provided significant habitat for wildlife.

Working with Ron Dixon, a graduate student at UMass, we took a quick look at subsamples of cemeteries in the Northeast and their wildlife. Dixon's master's thesis was converted to an article titled "Cemetery Ecology," which appeared in *Natural History*. The article was quickly republished in several trade journals for those concerned with the management of cemeteries. (Unfortunately, the names of the authors were reversed in the publication, and my name appeared first when it should have been Ron's.)

These initial efforts helped establish a new focus for a rapidly growing branch of wildlife biology and forestry. Given today's growing acceptance of wildlife management in cities and blooming urban forestry efforts, it is a pleasure for me to stare into the fire on winter evenings and remember our research group pondering the question just what the heck is urban forestry and wildlife? and what do we do now?

DeGraaf, Brush, Gordon, Moore, and I took full advantage of our association with UMass to earn doctoral degrees from that fine institution. Brush became a professor at the University of Wisconsin at Stevens Point, Payne went on to a sterling USFS career in international forestry in Paris, France, and Washington, D.C., and Moore continued his research career with the USFS. DeGraaf pulled off a trick almost unheard of in the USFS by spending his entire career in one location, progressing from technician to a top-level scientist and project leader. Maybe he was the biggest winner. If so, he deserved and earned it.

CHAPTER 4

A NEW PARADIGM, ECOSYSTEM MANAGEMENT:
LA GRANDE, OREGON, 1973-1993

In 1973 I accepted my dream job in La Grande, Oregon, where I spent the next twenty years as director of the USFS's Range and Wildlife Habitat Laboratory. In the early 1990s, I was assigned leadership of several teams composed of state and federal agency scientists and academics dealing with the implications of the Endangered Species Act—specifically the cases of the northern spotted owl, marbled murrelet, and salmon—in managing old-growth forests in the Pacific Northwest. Those efforts garnered both accolades and condemnation, including death threats that were taken seriously.

BIG ISSUES GROW FROM SMALL SEEDS

The purpose of my transfer to Oregon in 1973 was to reorganize the USFS's RWU at La Grande. During the previous decade, relationships between the RWU and the U.S. Fish and Wildlife Service's RWU at Oregon State University had withered, primarily due to lack of attention from the USFS, which was focused primarily on range issues. One of my assigned duties was to revitalize that relationship. To help me along in that effort, my immediate supervisor, assistant research station director, Robert Tarrant, allocated me $10,000 in additional funding for cooperative research with the Oregon co-op unit.

I called the co-op leader, Howard Wight, and asked him to put together three Master of Science–level research proposals for my consideration. A month later I journeyed to Corvallis to review the proposals. All three were well prepared, nicely presented, and worthwhile. So I told Wight that I had no preference and it was his call. He selected a proposal from Eric Forsman on the habitat preferences of the northern spotted owl, a little-known subspecies at that time. The work was to be overseen by the co-op's assistant leader, Dr. E. Charles Meslow.

Forsman's master's thesis strongly associated northern spotted owls with old-growth forests in western Oregon. Sensing that this finding might have a bearing on the developing political uproar over the cutting rates of old-growth timber on federal lands in western Oregon, Washington, and northern California, I arranged for additional USFS funding (approximately $20,000) to determine whether there was significant use of younger stands by northern spotted owls.

That study substantiated the original finding that the subspecies was significantly associated with old-growth forests, especially for nesting and brood-raising. It seemed highly likely that the northern spotted owl would be a candidate for being listed as either threatened or endangered under the auspices of the Endangered Species Act (ESA). The potential impact of Forsman's findings on timber yields from continued cutting of old-growth forests, especially on federal lands, triggered a number of other studies related to the northern spotted owl's preferred habitat. Such a reaction to any research finding apt to dramatically upset the status quo would likely engender, reasonably enough, more and more intensive research by an array of research teams.

Potentially, many millions of dollars and thousands of "good-paying jobs" were potentially at stake. I could not help but remember an admonition from an economics professor: "Remember, money talks and bullshit walks."

The cumulative impact of that research was manifested several years later with the U.S. Fish and Wildlife Service's finding that the northern spotted owl was indeed threatened under the auspices of the ESA. Adherence to the act would surely and dramatically have a negative impact on the timber industry, jobs, returns to the Treasury, and revenues to affected counties.

That, in turn, led to four separate efforts (I was part of all four and served as team leader of three) to come up with an appropriate management plan that would be deemed acceptable by the U.S. Fish and Wildlife Service and then the federal courts. The end result, the Northwest Forest Plan, which included a dramatic reduction in timber yields from federal lands in western Washington, Oregon, and northern California, withstood numerous determined attacks over two decades.

This was a dramatic fork in the trail for the management of federal lands in the Pacific Northwest and for many of the scientists involved—certainly including me.

Another related issue was bubbling on the back burner. One purpose of the Endangered Species Act of 1973 was "to provide a means whereby the ecosystems upon which endangered species and threatened species may be conserved." The emphasis on ecosystems was new and, oddly enough, was approached through "indicator species" that were determined to be threatened or endangered by the U.S. Fish and Wildlife Service. That led the USFS chief, F. Dale Robertson, to declare that henceforth the national forests would be managed under a new paradigm: ecosystem management.

But as the four supervisors of the three national forests in the Blue Mountains asked me, "What the hell does that mean?" They were not alone in their puzzlement. They requested that I put together a multiagency team to develop an approach to dealing simultaneously with all species of vertebrate wildlife in land-use planning. The effort was to be specific to the Blue Mountains but developed in such a manner that it could serve to guide in replicating efforts for other ecosystems across North America.

All 378 vertebrate species known to occur in the Blue Mountains of Oregon and Washington were categorized by their association with plant communities and successional/structural stages within those communities for purposes of reproduction and feeding. Special habitat features were evaluated, including riparian zones; edges; snags (standing dead trees); dead and down woody material; and cliffs, caves, and talus. It was recognized that managers would give some species more emphasis than others—in this case, mule deer and elk. Silvicultural options to obtain mixed objectives were developed and discussed. Determining impacts of various approaches to meeting wildlife habitat objectives were outlined. Seventeen authors were involved. The results were published by the USFS in 1979 as *Wildlife Habitats in Managed Forests: The Blue Mountains of Oregon and Washington.*

Taken together, ecosystem management and guidance on how to approach that job, so far as vertebrate wildlife species were concerned, was a significant fork in the trail for federal land management agencies.

MANNA FROM HEAVEN

In the winter of 1979, I received a phone call from Dr. Laurence "Larry" Jahn, vice president of the Wildlife Management Institute (WMI) in Washington, D.C. The time had come for WMI to upgrade several of its classic publications on the primary big game species of North America. That included Olaus J. Murie's *Elk of North America*, published in 1951. He offered me the honor of serving as the editor as well as the author of several chapters for a new book to be titled *North American Elk: Ecology and Management.*

After checking with my bosses, I quickly agreed, provided that I could bring in an assistant to do the detailed compilation and editing that would be required. I had just the man in mind: Dale E. Toweill, who had recently finished a master's degree at Oregon State University and was working for the USFS on a temporary appointment with me in La Grande. I did not yet consider myself an expert on elk, as my research concerning that species was just beginning. Twenty-five experts with appropriate *bona fides* were recruited to write the fifteen chapters. Richard E. McCabe and Larry Jahn of the WMI would serve as technical editors. Published in 1982, the book received awards from several organizations, including The Wildlife Society.

Eighteen years later, in 2000, Dr. Jahn, president of the WMI, invited me to put together a plan for and then compile and edit an updated and much expanded edition of *North American Elk: Ecology and Management.* I was to recruit the appropriate authors and convince them to volunteer their efforts.

Fortunately, Dale Toweill was just finishing his doctorate at Oregon State University and was available and eager to serve as my partner in this new endeavor. This effort involved thirty-nine authors and nineteen chapters. Richard McCabe of the WMI served admirably as technical editor. The book was published by the Smithsonian Institution Press in 2002.

During the course of putting the book together, I was being sucked more and more into the vortex of dealing with the northern spotted owl and old-growth forest management issue. Toweill moved more and more into the lead, so much so that I concluded that he should be given first billing as compiler/editor. After all, fair is only fair. It pained me but it was the right thing to do. Toweill, as I expected, performed above and beyond the call of duty.

By now, my efforts in research on the interactions between elk, mule deer, cattle, timber management, roads, and traffic were being turned over to my associates, Dr. Larry "Bear" Bryant and Dr. Michael Wisdom. It was a dramatic fork in the trail. Certainly, such was not of my choosing.

OLD SANCHO'S GHOST

Some complete stories emerge in parts, sometimes separated by decades. In the early 1960s, our bosses in the TGD assigned my partner Rod Marburger and me to work with two veterinary pathologists from the School of Veterinary Medicine at Texas A&M—Dr. R. M. "Mick" Robinson and Dr. D. O. N. "Don't" Taylor. At the time, much less was known about diseases of wildlife than is true today.

Rod and I had very little formal training in diseases of wildlife. What we did know we had picked up from our veterinary colleagues, assigned readings, and frequent references to the *Merck Veterinary Manual—Second Edition*. We paid for a lot of long distance telephone charges talking our way through problems with our veterinary colleagues.

The wildlife biologists on our informal team served as the point of first contact with ranchers, observing, taking detailed notes, conducting on animals suspected of disease, and preserving tissues for later assessment in the Veterinary Pathology Laboratory at Texas A&M.

The TGD's Black Gap Wildlife Management Area adjoined Big Bend National Park in far West Texas and was the location of an innovative, high-cost, high-risk attempt to return desert bighorn sheep to the Davis Mountains. Desert bighorns had been extirpated in West Texas in the early part of the twentieth century by a combination of overhunting and impacts of diseases and parasites contracted from domestic sheep.

As part of this effort, two square miles of the Black Gap were enclosed behind an eight-foot-high net-wire fence. The fence was intended to hold in bighorns while excluding mule deer, pronghorn antelope, cattle, and domestic livestock, which might compete with the bighorns for forage or expose them to disease. Electrically charged wires strung along the bottom and top of the fence were intended to keep predators, especially cougars, from entering the enclosure.

The desert bighorns inside the fence had been trapped in Arizona and were being held in the holding pen until they acclimated to the habitat conditions and their numbers increased to a level where release into the wild was deemed feasible. The Black Gap and the adjacent national park and the presence of personnel from those agencies provided some extra measure of protection from both poaching and exposure to domestic sheep.

The first group of desert bighorns transferred into the enclosure included a large, full-curl-horned ram christened "Old Sancho" in spite of *de rigueur* protocols of not naming animals involved in a study or experiment. It was considered more "scientific" to refer to animals by letters or numbers or some combination thereof. The rationale was that naming an animal might lead researchers to develop affection or affinity for that animal. That, in turn, might interfere with objectivity or, heaven forbid, introduce anthropomorphism into the research game. It was a charade, at least for the wildlife biologists involved. We felt what we felt regardless of protocol. But we played the game.

For the first several years, the reintroduction effort went according to plan as reproduction and survival rates of the bighorn population in the enclosure exceeded expectations. Things were going so well that the planned releases into the wild were twice delayed; after all, why not keep a good thing going? As time passed, Old Sancho became a truly magnificent ram.

Very late one night in mid-summer of 1966, I got a call at home in Llano from Tommy Hailey, a classmate in the class of '57 from Texas A&M and the TGD biologist in charge of game management operations in Trans-Pecos Texas. As part of his "duties as assigned," he was the manager of the Black Gap. He had discovered several bighorns dead in the enclosure and others showing signs of distress. He was looking for help—*muy pronto*.

I called Mick Robinson and Rod Marburger to meet me in Llano ASAP and geared up for a trip to the Black Gap to see what we could do to help Tommy with his problems. In the meantime, Mick instructed Tommy to keep an eye on the sheep in the holding pen and try to capture any that showed signs of distress.

We were crossing the Pecos River when we received a relayed two-way radio message that Tommy and crew had captured three distressed bighorns: two adult females and Old Sancho. All were incapacitated but still alive. Tommy and the three bighorns were en route to meet us at a roadside park just outside of Sanderson.

We reached the rendezvous first, and Tommy arrived less than an hour later. Mick gave the animals a quick examination. The linings of their mouths and their tongues were swollen and bluish purple. Even Rod and I knew what that meant but hoped we were wrong. We looked to Mick for confirmation. He said, "It looks like epizootic hemorrhagic disease—blue tongue." He asked Tommy, "Have the bighorns been in contact with domestic sheep?"

Tommy shook his head, "I don't know of any domestic sheep within miles of the wild sheep pens."

Mick speculated, "The most common carriers for blue tongue are *Culicoides* gnats, which are commonly associated with domestic sheep, especially around water sources. The gnats can move a couple of miles if aided by the wind. I suspect there are domestic sheep somewhere within a mile or two of the wild sheep pens."

Tommy asked Mick about the best move possible to protect the rest of the captive bighorns. I think we all knew that it was likely too late to protect the remaining wild sheep from exposure. We could only hope that some of the bighorns would not be exposed, or that some of the infected sheep would survive and thereafter be immune.

Neither outcome seemed likely. We moved the sick sheep from Tommy's truck to Mick's station wagon. One ewe died during that process. Mick performed a necropsy, right there beside the road, using the dropped tailgate of Tommy's truck for an operating table. Examination of the internal organs backed up his original suspicions: massive bleeding into the muscles and internal organs. Mick immediately put the surviving bighorns on an intravenous drip of saline solution to counteract dehydration. Tears ran Tommy's face as we loaded Old Sancho into the back of Mick's station wagon. Mick held Old Sancho's head in his lap so he could keep his nostrils cleared of mucous and aid his labored breathing.

Mick and I took turns driving through the night. Just before we reached the School of Veterinary Medicine at Texas A&M, the second female expired. Old Sancho breathed his last just before noon. Necropsies and laboratory analysis verified Mick's original diagnosis.

The next day Tommy circled above the sheep pen in a small aircraft and spotted domestic sheep on a private ranch adjacent to the Black Gap Wildlife Management Area. They had been brought in with no warning to the TGD. Of course, the rancher was not required to inform Tommy of his actions. And, to be fair, it was likely that the rancher had no idea of the danger his sheep might pose for the bighorns. Or maybe he just didn't care. It was a fiasco.

This was a huge setback to a noble and very expensive attempt to restore desert bighorns to the mountains of far West Texas. Live and learn was at the core of wildlife management in those early days and still is—perhaps much more so than we would like to admit.

In 1973, when I took over leadership of the RWU headquartered in La Grande, I soon became acquainted with the regional director for the Oregon Department of Fish and Wildlife, Will H. "Bill" Brown.

My predecessors had concentrated primarily on research related to livestock grazing and range management. When I arrived at my new assignment, Bill knew me as a wildlife biologist by reputation and through my publications and hoped that I would shift my RWU's emphasis more toward wildlife studies. He worked diligently to focus my attention on issues that he saw as "sore spots" in the USFS's primary focus. The current timber harvest practices, with rates of exploitation that he considered excessive, had long been a burr under his saddle.

Brown had cultivated a relationship with John Forsman, the deputy regional forester for the USFS's Pacific Northwest region who had jurisdiction over range and wildlife issues. Not long after my arrival in northeast Oregon, Brown invited Forsman and me on a week-long pack trip into the vast wilderness of the Snake River Canyon. The Snake River forms the boundary between Oregon and Idaho, and the area is, appropriately enough in my mind, called Hells Canyon.

The trip was set for mid-September, which was no accident of timing on Brown's part. Fall marked the end of the grazing season for domestic livestock permitted as part of livestock grazing allotments to private ranchers. Now, at the tag end of the sheep-grazing season, we encountered deteriorated range conditions that Brown and I considered ecologically and, in the long term, politically unacceptable. Many decades of overgrazing by domestic sheep had transformed an ecosystem dominated by perennial native bunchgrasses into one dominated by annual cheatgrass, an exotic invader from the Middle East.

The sheep bed-grounds we encountered, where sheep congregated to spend the night while being watched over by herders and sheep dogs, had been denuded of vegetation. In some such places, some of the topsoil had blown away. Those who rode drag when the sheep were being moved commonly had bandannas tied over noses and mouths to filter out the fine volcanic ash dust raised by the hooves of the sheep and horses.

There was no mystery as to the cure for the problem—alterations in the grazing program, including a reduction in the numbers of sheep allotments and better control of those that remained. Bill made no mention of the sorry state of range conditions; he didn't need to, as neither John Forsman nor I were blind to the implications of what we seeing.

Upon his return to the USFS regional headquarters in Portland, Forsman went to work on the problem. Within a decade, a number of domestic sheep allotments in the Snake River Canyon west of the Snake River were replaced by cattle allotments. Some allotments were put into nongrazing status pending improvement in range conditions. Within a decade, range conditions were improving in the upper regions of Hells Canyon, especially on the "benches" halfway up the slope from the Snake River.

Then the Oregon Department of Fish and Wildlife (ODF&W), in cooperation with the USFS, believed it was time for reintroduction of bighorn sheep into Hells Canyon. By the early 1990s, bighorns could be seen grazing the canyon's slopes, a testimony to federal-state cooperation and the leadership of wildlife biologist Vic Coggins and associates from the ODF&W.

By the time I became chief of the USFS in 1993, most, but not all, of the domestic sheep had been removed from national forests in Hells Canyon. The remaining permit holders for sheep grazing on the Idaho side of the Snake River Canyon had considerable "pull" with the Idaho governor, state legislators, and Idaho's congressional delegation. Those forces successfully thwarted USFS efforts to move the sheep-grazing operations to allotments outside Hells Canyon.

Then, as long feared, blue tongue disease showed up in the bighorn sheep. But unlike the situation decades earlier in Texas, the USFS, working closely with the Idaho Fish and Game Department and the Oregon Department of Fish and Wildlife, had authority to develop and institute actions favorable to the wild sheep. The only thing lacking was collective gumption forcefully applied. Robert "Bob" Richmond, supervisor of the Wallowa-Whitman National Forest, which lies on the west side of the Snake River Canyon, had the lead. Understandably, he moved cautiously and slowly, which was frustrating to some. An old hand, he fully appreciated the role of politics in public land management.

In the midst of the dustup, Bill Brown, now long retired from ODF&W, arranged a horse pack trip for Vic and me into the Snake River Canyon. Vic, the wildlife biologist who was considered the "daddy" of the wild sheep reintroduction into the canyon, was fiercely proud and protective of "his" bighorns. During the trip, he lobbied us constantly about what he considered "a too damned slow" pace in removing the last bands of domestic sheep from Hells Canyon. He convinced me that it was impossible to establish and maintain enough separation between domestic and wild sheep to preclude transmission of disease (including blue tongue) from domestic to wild sheep.

That first night, I lay in my sleeping bag studying the stars and pondering the picture that Vic had painted. A vision bubbled up from the depths of memory—Old Sancho, trussed and blindfolded, struggling to breathe with his head resting in Mick Robinson's lap during the all-night trip from West Texas to the veterinary school at Texas A&M. The vision was vivid, disturbing, and—too likely—prophetic.

The day we reemerged into civilization, I stopped at a USFS ranger station and called regional forester John Lowe and Wallowa-Whitman National Forest supervisor Bob Richmond, both old friends and respected colleagues. I encouraged them to move ahead as aggressively as possible with the removal of domestic sheep allotments from Hells Canyon.

John Forsman was well equipped to take the lead. The last permit holder's grazing operation was to be relocated to a vacant allotment—if at all possible, to a better allotment than he currently held. The permit holder was to be made whole and happy or more. All flak from elected officials was to be directed to me.

I was quickly and bluntly informed by the regional forester and the forest supervisors involved that they would take the lead. They would do their jobs, and I could do mine. This was their job, their territory, and they would take care of it. Though I had expected no less, I was glad to hear their response. Lowe and Richmond moved ahead with political sensitivity and skill. A potentially explosive task was skillfully handled.

Several years later, after my three-year tour as USFS chief was over and I was teaching school at the University of Montana, I took a jet-boat excursion with my new bride up the Snake River from Lewiston, Idaho. Midway in the trip, the skipper pointed out a full-curl bighorn ram standing on a rock outcrop far above the river's edge. He said that the ram was often spotted along this stretch of the river and that the biologists called him "Idaho 33." Somehow, I preferred to

John Forsman (left) and Jack with the day's take during a week-long grouse-hunting trip into the Snake River country of Oregon's Hells Canyon.

think of him as Old Sancho, not really giving a damn at this point if that was anthropomorphic.

As I write this, I can look up at the wall of my home office and see a marble-on-wood plaque known as the Federal Statesman Award, presented to me by the Foundation for North American Wild Sheep in 1997 in appreciation for my "lifetime commitment" and "true dedication and support of all wildlife, including wild sheep." The plaque's engraving of a bighorn ram with full-curl horns looks very much as I remembered Old Sancho. Fair enough, for when I accepted that award, I had Old Sancho on my mind.

"BUT ARE YOU *REALLY* HURT?"

When I was the leader for the RWU at La Grande, my personal research focused on the interrelationships of Rocky Mountain mule deer, Rocky Mountain elk, and cattle relative to forest management. My closest associate in this endeavor was Dr. Larry "Bear" Bryant, a onetime paratrooper and college basketball player. He stood about six feet three inches tall, had a full head of black hair, and sported a full, carefully trimmed black beard increasingly flecked with gray. He held three academic degrees in wildlife biology and range management, including a Ph.D. Bear was blessed with considerable charm—which he could turn off and on at will without missing a beat.

By the 1980s and early 1990s, USFS land managers were deeply involved in land-use planning and desperately needed more detailed knowledge of how deer, elk, and cattle interacted and reacted to varying habitat conditions, including the presence of roads, traffic on roads, logging, hunting, and livestock grazing. What would happen when additional vast reaches of national forest were brought under intensive forest management and mule deer and elk had to deal with those conditions? Larry and I concluded that a broad-scale yet tightly controlled experiment was essential to providing defensible answers to that critically important, politically hot, and economically significant management question. In order to do that, we needed a study area of adequate size with confined populations of mule deer, elk, and cattle where we could compel the mixed bag of ungulates to react to imposed changes in habitat conditions as opposed to their simply moving away from those conditions.

We proposed to enclose some forty square miles of the Starkey Experimental Forest in the Wallowa-Whitman National Forest within a fence that would contain deer, elk, and cattle. Then we planned to develop and install a fully automated radio-telemetry system to systematically, accurately, and frequently locate the deer, elk, and cattle under study.

To carry out our proposed study, which came to be called the Starkey Project, we needed to be able to repeatedly capture and handle deer, elk, and cattle to install and service the tracking collars, judge body conditions, ascertain pregnancy rates and age, and draw blood samples. As Starkey was spring/summer/fall range, it would be necessary to concentrate the deer and elk during the winter when they would have migrated to their winter ranges at lower elevations and cattle would have been removed to lower elevations on private lands. Deer and elk were artificially fed a diet with a caloric intake that simulated their diet on the winter ranges to which they had previously migrated.

Nobody, nowhere, nohow, at no time had ever managed to pull off anything quite like we had in mind. We figured that it would cost, up front, some $2 million, which, of course, we didn't have. Building support and public acceptance for a forty-square-mile elk and deer enclosure in the middle of the Wallowa-Whitman National Forest would require powerful salesmanship coupled with a tall order of politicking inside federal, state, and county governments, in the ranching community (especially among the permittees who ran cattle on Starkey), and among hunters and the general public.

In our research world, the first step we faced in selling a new, expensive (some said radical) idea was to get the right people camped out on the ground for several days in order to create a captive audience in an appropriate setting. Commonly, we stayed in the USFS cabins at Starkey, cooked outside, and had plenty of time for discussions and explanations—with maybe a drink or two for those so inclined. In the process we eventually got around to our well-prepared and carefully staged sales pitch.

Drinks and cooking over an open fire gave us to time to present our vision. The same trips could have been taken via pickup trucks in a third of the time. But the horses and the camp-out summoned up the *mojo* of the old-time USFS and gave us more time and opportunity to stop at all the right stops and make our pitch and answer questions.

With some groups of VIPs, we ended up with a few more folks than we had horses at our disposal. So we sometimes had to borrow or rent enough horses to get the job done. Bill Brown was good about giving me access to his personal string of horses. Sometimes we had to combine several strings of horses that were strangers to one another.

Horses, much like people, have best buddies, social groups, and pecking orders. When horses that are strangers to one another are thrown together, a process of "getting acquainted" or "sorting things out" begins immediately and continues sporadically, regardless of the riders. Such is a horse thing unrelated to the riders. To minimize the opportunities for horse wrecks, it is wise to keep horses separated into their accustomed social groups.

On one such occasion, we mounted up in the headquarters compound with the horses and their riders carefully separated into two groups of equine *compadres*. I was riding my big, tall, thoroughbred sorrel mare named Casey, who was usually employed as a packhorse. When I stepped up, she crow-hopped around with her head between her knees. Then she commenced bucking in earnest. When her little snit was over, I was well aware, from past experience, that Casey hadn't really put her heart into it.

In the late afternoon, we were headed down an old logging road back to the Starkey headquarters. The assistant deputy chief for research from the USFS's Washington office, Robert "Bob" Harris, who had once held the position I now held, was mounted on a big black gelding from the other horse clan. Without thinking, he broke the order of march and rode up beside me. His gelding and Casey laid their ears back and tossed their heads. Then his gelding bit Casey in the neck.

Casey planted her front feet and came to a sudden stop, jerked her head up, and shied, throwing me forward in the saddle. My nose and the poll of her rather hard head collided, knocking me back in the saddle. I saw stars and grabbed on hard with my legs in an attempt to keep my seat. Unfortunately, my right foot, with spur attached, was out of the stirrup and caught Casey hard in a tender spot. She commenced bucking—big-time. I contemplated the saddle horn on my way down as my ass went higher than my head.

The logging road had a steep shoulder that sloped down some twenty feet to the creek bottom. I landed on that slope, which knocked the wind out of me. When I quit bouncing, I was upside down in the dry creek bed. I managed to wiggle around, got my head higher than my butt, rolled over, and sat up. Choking on blood, I rocked back and forth with my head in my hands, sucking hard and trying to get my wind back. From the blood and tears pouring down my cheeks I couldn't tell if I was really hurt.

Bear jumped off his mare, took a flying leap off the side of the logging road, and slid down the bank to where I sat, rocking back and forth and holding my head in my hands. When Bear pulled my hands away from my face, he could tell that my nose was pushed over to one side and there was a significant L-shaped gash on the bridge of my nose. In a solicitous tone, he commented, "Geez man! That's gnarly! Are you hurt?"

"Gnarly? Am I hurt? You silly bastard! Am I hurt!?"

"Well, hell, I can see she broke your nose. But are you *really* hurt?"

Most of the wildlife biologists and foresters that I worked with over many years could be labeled as pretty damn tough. Simply "cowboying up" was the expected way to deal with pain resulting from any injury that was not immediately fatal or immobilizing. Given my audience, I figured it was time for me to cowboy up, crawl up the bank to the road, and "get back on the horse." After all, reputations were at stake. Admittedly, I was having a little trouble focusing both eyes on the same thing at the same time. With Bear pushing from behind, I crawled up the bank to the road, stood up, and gathered up my horses' reins. I was dizzy, a bit confused, and getting a little sick to my stomach. I tried twice to mount up and didn't make it.

It came to me that the better part of valor would be to lead Casey—who was now, to my mind, the bitch from hell—the remaining half mile to headquarters. She was wild-eyed and dancing around. I thought, to hell with the "cowboying up" bullshit. Truth be told, I would have flopped down flat on my ass, held my face in both hands, and cried if I had thought it would help.

I stepped up into the saddle and got a good seat. Casey rolled her eyes, but she was through bucking. By the time we got to camp, my eyes were near to swelling shut. One look in the rearview mirror of the horse truck told me that I needed some stitches and some expert "mashing" of my nose into some resemblance of its original shape. After a couple of drinks of 90-proof bourbon straight from the bottle—this time for actual and legitimate purposes—I drove the thirty-five miles into La Grande to the emergency room at the hospital. I knew the doc on duty from Lion's Club. I should have lied when he asked what happened. After he quit laughing, he did a nice job of rearranging my nose before sewing it up with fourteen fine stitches.

By the time I got back to Starkey, a party had broken out; after all, the show must go on. Bear was grilling steaks and convincing his audience of the need for the research facility we had in mind. My front teeth were a little loose, and my upper lip was joining my eyes and nose in puffing up and turning delicate shades of red, blue, and purple. I couldn't breathe through my nose, as it was packed with cotton to help hold its shape and stanch the bleeding. Salving my wounds by sipping whiskey and chewing aspirin seemed vastly preferable to trying to chew steak. Just after supper, modestly fortified against pain and the dangers of infection, I crashed in my bunk.

The next morning, after we saddled up and after everybody else was mounted, I led Casey across the road that ran in front of the Starkey headquarters and down into a muddy bog below the fenced-in spring from which we pumped our water. I hoped the bog would dampen Casey's bucking abilities. Besides, if Casey won the argument, it would be better to land in the mud rather than on rocky ground.

It was, however, essential that Casey regain some understanding of who was boss. I pulled her head down to my left stirrup, stepped up, and swung into the saddle as she spun left. I got my right foot in the other stirrup before she started bucking—a good thing! Casey bucked her best—and a damn good job it was, considering that she was nearly up to her hocks in mud. Her every jump, both the going up and coming down, made me keenly aware of the bulbous condition of my nose, my loose front teeth, and why I had downed a double snort of bourbon and chewed a half-dozen aspirin tablets for breakfast. I bucked Casey around and around in the muck until she stood with her nose between her knees slobbering and blowing hard.

I spurred her out of the mud, then dismounted and mounted a half-dozen times as she stood rock solid. Lesson learned. By now, it was even money as to which one of us was the most bucked out.

For the rest of a long day, all the horses, including Casey, behaved perfectly. As we bid our guests farewell the next morning, several commented on the lack of concern Bear had displayed when he dismissed my horse wreck and my broken nose. He reminded us that after seeing my head and Casey's collide and watching as I fell ten feet or so to land on my head and shoulders, he had good reason to believe I had broken my neck or cracked my skull. Under those circumstances, a broken, split-open nose could be reasonably dismissed as a trifle.

The most important result was that we sold our guests on the new research effort. One guest confided to me, many years later, that he figured if Bear and I could stay on message despite a horse wreck, broken nose, stitches, eyes nearly swollen shut, teeth loose, and being a little drunk, then he couldn't help but be supportive. But believe me, that was one tough way to get support for a research program!

KILL A COMMIE FOR CHRIST

The Starkey Project was the realization of an impossible dream. First, we had to build the elk- and deer-proof fence. USFS engineers told us it would cost three to four times what we had anticipated. Costs like that would kill the project—even a big dreamer would choke on the required investment. Solution? Forget our dream or find a way to get the costs down dramatically.

So we finagled getting Bear to New Zealand for a couple of weeks to check out the fencing that we wanted to use. The fencing had been developed by the Kiwis for use in their rapidly developing industry of raising of red deer—the same species as elk but a different subspecies—for production of meat, primarily for export. He charmed his hosts into selling us the special wire for cost and, on top of that, to split the costs of shipping. Of course, they had an ulterior motive in that they wanted to get a New Zealand product into North America for potential use by the rapidly developing game farming market.

Now we needed bids to build the fence. Fortunately, we had a supporter in Gary Price. Gary was in his early sixties. His face reflected decades of operating logging and road-building equipment in all kinds of weather. A former football player at the University of Washington, Gary held a degree in forestry and had been a successful "gyppo" (contract) logger. He had had his disputes with the USFS timber sale administrators over the years. On top of that, he reveled in his quite conservative political views. After doing some "figuring," he reckoned that the USFS engineers were way off the mark relative to costs and that he could build the fence for the budget we had developed. The engineers retorted that Gary and Bear, not being engineers, were way off base.

Hundreds of twelve-foot treated fence posts had to be anchored into postholes drilled into solid limestone in some places. Hey, not too worry. Gary invented a newfangled drilling machine mounted on a bull-dozer-tracked vehicle that could traverse steep and rugged terrain with a drill head that could be tilted 180 degrees to assure that the posthole was plumb. That eliminated having to level the fence right-of-way.

And then he came with a perfectly logical question that had not been considered. Where does it say in the Bible, or even the USFS manual for that matter, that fences have to be built in straight lines or right on an internal property line? He figured that if he could take advantage of the terrain and build a "wiggly" fence, he could save "about a ton, maybe a ton and a half" of money. Starkey was enclosed on three sides within the Wallowa-Whitman National Forest and all on the La Grande Ranger District. So did it really matter if the fence wiggled around a little as long as the acreage—inside and outside the fence—came out right?

I proposed a "wiggly" fence line to district ranger Robert "Bob" Shrenk, who took pride in being a tad unconventional. He saw no problem. We kept the agreement to ourselves.

Jack and Larry "Bear" Bryant (right) *working out the logistics of a groundbreaking proposal to create a 25,000-acre enclosed landscape for applied research, which came to be known as the Starkey Project.*

Gary was already drilling postholes about three times faster than the USFS engineers had thought possible and at a third of the cost projected. Gary got the fence built ahead of the agreed-upon completion date and right on his estimated cost. Thinking and acting "outside the box" can pay dividends. The fence gods must have smiled—or maybe laughed out loud. But "we got 'er done!"

We thought we had taken care of most of the public relations problems associated with our project; we had certainly put enough effort into it. But some folks still had concerns, and a few remained bitterly opposed. There was the question of who would be allowed to hunt inside the fence and how those hunters would be selected. We proposed a special drawing for permits, and the ODF&G agreed.

Sadly, folks who had hunted on Starkey, some for generations, were displaced unless they drew a permit to hunt inside the fence. Their irritation and sense of loss were understandable. Furthermore, the fenced-in area sat in the middle of an elk and deer migration route between summer and winter range. While we had confidence that the elk and deer would simply go around the fenced-in area, some very vocal critics thought that they would simply stack up against the fence and die. Finally, some didn't understand that their treasured rights of access for recreation or collecting firewood or using the land in varied other ways would be maintained. Starkey was, in all respects, to remain public land. Yet there was, reasonably enough, concern and suspicion in some quarters.

So we went back to supporters in the national forest system for help. The regional forester agreed to "partner up" and employ a seasoned wildlife biologist to formulate and execute a public relations plan. He would present programs and guide tours for any and all interested in the project. Senior wildlife biologist Michael "Mike" Wisdom transferred over from the BLM to fill that role.

Over the next three years, he conducted hundreds of tours of Starkey; appeared routinely on radio and television; wrote articles for newspapers and magazines; held public meetings around the state; and addressed dozens of service clubs, hunting organizations, and anyone else he could get to sit or stand still and listen. He proved a master at spreading oil—some called it snake oil—on troubled waters.

Wisdom's efforts freed up Bear and me and our ODF&G partners, Donavin Leckenby and Leonard Erickson, for our research duties. With my encouragement and support, Wisdom attended the University of Idaho during winter semesters and eventually completed his doctorate. When I became the USFS chief in 1993, he would replace me as the project leader for the Starkey Project.

I was along on a tour that Wisdom conducted when the caravan of vehicles proceeded down the main road until they came to the first of a series of big ponderosa pine trees that had posters nailed to them about twelve feet above the ground. The signs proclaimed, in various ways: "JACK WARD THOMAS, COMMUNIST AMERICAN." Several had my image fixed squarely in the cross hairs of a rifle scope. One of the more artistic had the red star of the Soviet Union in a circle with cross hairs pointing out the center of the star.

Around the circle was script in Russian that, upon translation, read "KILL A COMMUNIST FOR CHRIST!" If the signs were meant to inflame the crowd against me or the Starkey Project, they failed miserably. In general, the public was disgusted or amused or both. Even one of our foremost detractors spoke out strongly in a letter to the editor of the local paper in opposition to what he had seen. As word of the signs spread, we were contacted by news outlets from across the country. A potential for a public relations coup had been laid in our laps, and we saw it and took full advantage.

A USFS law enforcement officer carefully removed the signs to be tested for fingerprints and potentially to be used in evidence. We found that the red star and surrounding wording was the identifying symbol for a local militia group that was reportedly arming themselves with appropriate weaponry and training to resist the Russian/Chinese troops they fully expected to invade the Pacific Northwest. That invasion was to come on the heels of the United States turning over sovereignty to the United Nations—or some such crock of B.S. Then, when that went down, the "militia" would "take out" local law officers and government officials who were known or suspected to be part of the conspiracy; blow up the bridges into the Blue Mountains to slow the advance of the Russian/Chinese armor; and retreat into the wilderness areas to carry on guerrilla warfare. Evidently Bear and I were high on their list.

It was not all that difficult to figure out who had put up the signs. This individual, who had been displaced from hunting on Starkey unless he drew a permit, was considered by some who knew him as "just a tad wacko." We had seen him slinking around Starkey dressed in military-style camouflage clothing carrying what appeared to be a Russian-style Kalashnikov AK-47 military assault rifle. On several occasions, crew members had heard what they thought were bursts of semiautomatic weapons fire. Was that our head case playing Rambo, or was he simply announcing his presence? Was he trying to be intimidating? Did he have a screw loose—or several? Or was it all of the above?

We informed USFS law enforcement officers, the FBI, and several county sheriffs. They correctly pointed out that we had not observed our suspect committing an illegal act. Running around in the woods packing an AK-47, dressed in camo clothing with one's face painted in green and black streaks, and "playing army" was a little weird, but not a violation of law. However, they would "keep an eye out." Okay, it was clear that, barring an overt assault, we were on our own.

Then, one fine day, I saw our "commando" coming down a trail, dressed in full military camouflage with his face daubed with camouflage paint and carrying an AK-47. He was playing soldier, but poorly. I could have whacked him with a rock—and was tempted to do so. I hid behind a large ponderosa pine next to the trail. When he was right on top of me, I stepped out. His AK-47 rifle was slung across his chest. He could not shift the weapon into firing position before I could grab him, and we both knew it.

I was at least six inches taller and nearly 100 pounds heavier. I said, "Howdy, Chucky"—using the diminutive of his name. "Chasing Commies?" His usually rather vacant eyes were flared open, and the proverbial cat seemed to have purloined his tongue.

"I know you put up those Kill a Commie for Christ signs. The cops—fed, state, and local—know it too. You are definitely on their radar."

I could tell that, close-up and face to face, he was harmless—even with his AK-47. He damned near wrecked himself scrambling back down the trail, looking back over his shoulder, and ricocheting off a couple of trees in the process. I hoped that he had pissed his pants.

I knew that Bear was in a local shooting club with Chucky and told him what had gone down. Sometime later, a little fairy whispered in my ear that Bear had engaged in a intense conversation with our potential bad boy. I was told it unfolded along theses lines. "Thomas and his family are very close friends of mine. I even like their wiener dog and yellow fuzzy pussy cat. It would upset me terribly if anything were to happen to any of them. If something did happen, I couldn't help but believe you were responsible. I really don't know how I would react. Understand?"

Our bad boy reportedly nodded, several times, and that was the end of that. We heard no more bursts of semiautomatic weapons fire at Starkey, and we never saw our "commando" in the woods again.

He must have decided to hunt down Commies or prepare to meet the pending Russian invasion somewhere else. There were no more signs nailed to the trees at Starkey.

Unfortunately, I was not able to stay with my dream research project until it was time for me to retire. I was shifted to other assignments—three sequential but related efforts—to deal with the burgeoning issue of sustaining threatened northern spotted owls and anadromous fish in the Pacific Northwest.

But our dream project—Bear's and mine—had come to be. We learned that standing up to attempts at intimidation had led to a new fork in the trail. Those who carried out the project likely did it as well as—or even better than—I would or could have done. Maybe they weren't the political operators that Bear and I had been, but they were damned fine researchers

Now, fast-forward' to June 23, 2012, to an event billed as "Celebrating 25 Years of Research and Partnership on the Starkey Project," sponsored by the Rocky Mountain Elk Foundation. Considering the trials and tribulations of carrying out our dream project, the celebration pleasured me greatly.

The Starkey Project had prospered against tough odds and produced a series of findings that dramatically altered the understanding of how elk, deer, and cattle are best considered in forest management. New habitat models developed through cooperative research with other agencies and the private sector, especially the National Council for Air and Stream Improvement (through Drs. Larry Irwin, John Cook, and Rachel Cook), received widespread praise.

It is rare when a unique, daring, outside-the-box, high-risk idea/dream actually comes to fruition. It is an even better day when that idea pays off beyond expectations. I was pleased to have lived long enough to say to naysayers, "We told you we could do it—*and we did!*" And the beat went on.

THE YELLOWSTONE FIRES OF 1988—HARBINGER OF CHANGE

The consequences of huge uncontrolled forest fires around Peshtigo, Wisconsin, in 1871 and Hinckley, Minnesota, in 1894 horrified the nation. Then the Big Burn of 1910 exploded in the northern Rockies,

Washington State, Minnesota, Wisconsin, and Yellowstone National Park. In Montana and Idaho more than a million acres of forested land burned in less than three days; eighty-five people (including seventy-eight firefighters) were killed. It was from those fires that emerged the culture of fire suppression that dominated federal land management policies for generations. The USFS's "10 a.m. policy" set a goal that a wildfire was to be extinguished by 10 a.m. on the day after discovery.

Sometimes wildfires creep along the ground killing some small trees and merely scorching larger trees. At the other extreme, a wildfire can literally explode through a forest, consuming nearly everything in its path. Such fires sound like the huffing and puffing of steam engines as they suck in oxygen and create their own winds to feed the flames. Once, in 1988, I watched from a helicopter as a runaway wildfire that had exploded in Yellowstone National Park brought "the wrath of God" to my mind.

In 1957, Starker Leopold, professor at the University of California at Berkeley (the son of Aldo Leopold, the "father of wildlife conservation") urged federal land managers to reconsider the role of fire in wilderness areas and national parks. His logic: because forests had evolved over the ages with fire as part of their ecology, the exclusion of fire was bound to have long-term, not yet fully understood, but most likely negative effects. For example, some seeds, such as those from a variety of lodgepole pine, open only after being exposed to the heat from fire. Dr. Leopold believed it was vital that we more completely understand the role of fires in ecosystem function and apply that knowledge in forest management, including the long-long term management of wildfires. Leopold's message set off a still ongoing and intensifying evaluation of national wildland fire policies.

In 1978 the USFS approved prescribed burning as a management technique and declared that "natural fires" would be allowed to burn—"under prescription"—in legally designated wilderness areas. That policy was based on two principles: (1) designated wilderness areas should be subject to the normal forces of nature, including naturally occurring fires; and (2) pragmatically, those areas were selected for the new approach because no timber was expected to be harvested from wilderness areas in any case. Therefore, there was no economic loss to be incurred and risk to firefighters was reduced.

That pronouncement did not change wildland firefighting as much in practice as in theory. The philosophy of immediate attack on wildfires, by and large, continued to prevail. For good reasons, federal land managers were fearful of the economic and political consequences of a "let burn" fire exploding into a large-scale wildfire that caused significant economic, social, and aesthetic losses, especially the loss of structures, livestock, and human life.

Those risks and fears haunted fire bosses who made the calls on whether to attempt to suppress wildfires or let them burn. These are among the biggest "damned if you do and damned if you don't" quandaries that face federal land managers. That seems apt to worsen as the effects of global warming become ever more manifest and fuels continue to accumulate in unmanaged forests due to fire exclusion.

Between 1972 and 1987, lighting strikes ignited some 235 natural fires in Yellowstone National Park that burned over some 34,000 acres. Initially, the "natural fire" policy was deemed to be working, and managers were encouraged.

Then in mid-June of 1988 the first major wildfire of the season in the Greater Yellowstone Ecosystem began in the Custer National Forest in the Absaroka–Beartooth Wilderness. It was treated as a natural fire and allowed to burn. Within a week, a lightning strike started a fire inside Yellowstone National Park at Shoshone Lake. A few days later, fires broke out in the adjacent Targhee National Forest and in the park's Fan Creek drainage. The wildfires in the national forests—outside of wilderness areas—were, in keeping with policy, controlled and then extinguished. The Fan Creek fire, in compliance with plans, was allowed to burn.

Then several other smaller fires started in Yellowstone and in adjacent national forests. The USFS fought the wildfires in the national forests, but the fires in Yellowstone were not suppressed. On July 11, a lightning strike started another fire in the Bridger-Teton National Forest that was allowed to burn. Then another fire started in Yellowstone south of Cooke City and was declared a "let burn" wildfire.

By late July, National Park Service and USFS managers were no longer making the calls—nature was. Now that private property and lives were threatened, "let it burn" was no longer a viable political or pragmatic management option. Suddenly the largest wildland firefighting operation in history was under way with all-out efforts at suppression by all federal land management agencies. State, county, and city resources were also brought to bear.

Every available resource was thrown into the fight: ground pounders, smokejumpers, mechanized equipment, and retardant-carrying fixed-wing aircraft and helicopters. The fires, in combination, continued to overwhelm suppression capabilities. At the same time, federal, state, and local political forces were demanding, ever more loudly, that the fires be extinguished. Those demands were much like standing at the edge of the ocean in hurricane winds and ordering the waves back.

By August 12, over 200,000 acres within Yellowstone National Park had been burned over. The fires—big news at first—were becoming old news as time passed.

Then, on August 20, all hell broke loose. By mid-morning, winds were blowing at fifteen to twenty miles per hour and gusting to seventy. The winds morphed into a steady gale that snapped off thousands of trees and produced a firestorm. In eight hours the acreage within the fire lines doubled to 480,215 acres.

Firefighters were now essentially powerless to affect the course of events. No agency spokespersons stood in front of microphones spinning the benefits of "natural fire" or "let it burn" policies. Senator Malcolm Wallop of Wyoming showed up and demanded the dismissal of the National Park Service director, William Penn Mott, who had defended what the press now pejoratively termed the "let it burn" policy. Senator Alan Simpson of Wyoming wanted the hide of Yellowstone's fire ecologist.

The overall implication of the press coverage was that wildlife—especially deer, elk and bears—were being destroyed en masse. In fact, except in the case of the most rapidly moving fires with dramatic "spotting" ahead of the fire, most of the wildlife species that were highly mobile moved ahead of or away from such fires, and most returned to the unburned potions of their home range within a short time. There were, of course, exceptions, and it was those exceptions that made the news. The public's image of wildlife and wildfire resembled the responses portrayed by Walt Disney is his late 1930s movie *Bambi*.

Fires do not destroy ecosystems; they simply change their status—temporarily. But this moment was not the time to try to teach politicians about ecological processes or confuse them with facts. Politicians were loud in their castigations and vigorously looked for scapegoats. They might have been wise to have all the facts at their disposal and to study them, at least briefly, before such posturing. But unfortunately such is rarely the way of politicians in the midst of crisis. Too many quick-on-the-trigger politicians have their next election in mind and seize opportunities to score political points. Of course, their asses are not on the fire lines, and they bear zero responsibility.

By late August, nearly 10,000 firefighters were on the fire lines in and around Yellowstone. On September 8, the area around Mammoth Hot Springs was evacuated. The evacuation had been delayed due to pressure from state and local officials concerned about further negative economic impacts on an already sorely diminished tourist season. Occasional wind gusts were recorded at over eighty miles an hour. Large pieces of burning wood called firebrands fell in the parking lots at park headquarters, where hundreds of firefighters had gathered to ride out the firestorm.

I was there after spending a week observing operations as part of a National Academy of Sciences review team. I had no fire shelter but thought I would be safe enough in the vast parking lot. When I heard the roar of the approaching flames and the firebrands began falling, my confidence in that assumption waned. The copper taste of adrenaline was strong in my mouth. I had never been more scared in my life—and that's saying something. There was no place to run and no place to hide.

There was no better option that I could determine than to lie down and curl up on the asphalt in the middle of the parking lot, assume a fetal position, hold on to my hard hat, and begin bargaining with the Lord God Almighty. At that moment, I understood that even to talk about putting out such a fire as the one raging around us was to engage in the most outrageous of fantasies.

When the winds abated and the smoke cleared, to my amazement, all the buildings around the parking lot were still standing and none were burning. Quick thinking by the fire bosses and precautionary "wetting down" of the beautiful wooden buildings had saved them. I felt like what I imagined Lazarus felt after rising from the dead. I was plenty scared for a while, but without a proverbial scratch.

Then, within forty-eight hours, the fires suddenly weakened as cold mists swept over the area. On September 11, the first snows arrived and ended the fire season—after some 1.2 million acres (1,875 square miles) had burned over.

The 1988 wildfires in the Greater Yellowstone Ecosystem brought the managers of federal lands and the concerned public to a grand fork in the trail. The belief that, with enough will, planning, troops, financing, equipment, and applied science, federal land management agencies could contain all wildfires had been, most emphatically, relegated to the realm of mythology.

The "10 a.m. fire policy" might have made perfectly good sense in 1910 when the trees in the national forests were considered stock in a warehouse to be protected until harvested, milled, sold, and used. The 1990s saw a dramatic collapse of these timber harvests, which further weakened the foundations of the old policy.

Clearly, federal land management agencies would be forced to reconsider wildland fire management policies. The National Academy of Sciences team, of which I was part, forcefully made that point. Over the next couple of years, things did change, but not as much and not as fast as our team would have liked.

These changes were rooted in the philosophies of an array of ecologist-philosophers manifested in such phrases as "the first rule of intelligent tinkering is to save all the pieces," "everything is connected to everything else and there is no free lunch," "ecosystems are not only more complex than we think—they are more complex than we can think," and "fire is a natural part of ecosystems."

This move to ecosystem management was preordained by the passage of the Endangered Species Act in 1973, whose stated purpose was "the conservation of ecosystems upon which threatened or endangered species may depend." The Yellowstone fires of 1988 dramatically impacted the evolution of new policies for the management of the federal lands relative to wildfire. But dramatic, terrible lessons still lay ahead.

THIS WAY TO THE REST OF YOUR LIFE

During the early 1990s, I was drafted to serve on three teams in quick succession to deal with perhaps the most contentious ecological, social, economic, and political natural resources issue relative to public land management faced by the USFS since the establishment of the national forests in 1905: how were the federal agencies to comply with the ESA in dealing with the issue of the northern spotted owl, which was listed as a threatened species and associated with old-growth forests on public lands east of the Cascades in Washington, Oregon, and northern California?

The national forests did not become a major source of timber for the country until after the end of World War II. At that time, timber available from private lands was believed to be inadequate to meet postwar demand. It was then that the nation turned to federal lands to supply the wood products needed to sustain the rapidly growing housing boom.

Steadily increasing harvests of old-growth trees kept lumber mills humming, workers employed, and the housing market supplied with quite reasonably priced wood products. That kept the Knutson-Vandenberg portion (10 percent) of timber sale receipts pouring into the treasuries of counties that included those national forests. The leaders of the USFS made it clear to successive administrations and Congresses that timber was being harvested from the national forests at unsustainable rates, given inadequate coincident investments in forest management.

That discrepancy was euphemistically referred to as "temporary departure from non-declining even flow" of timber to markets. As "non-declining even flow" was required by law, it was reasonable to believe that "temporary departure" was really no departure at all and would be made up later. And, after all, politicians promised that this departure would be compensated for in two ways. First, the amount of timber cut from national forests would be reduced as the supply from cut-over—now regenerated—private lands increased over time.

And increased federal funds would be made available to the USFS to employ the latest silvicultural techniques to make up the difference over time.

Surprise! Surprise! The full funding for intensive timber management and road management never came, so it was inevitable that timber yields from the national forests ratcheted downward and then essentially collapsed in the early 1990s. That collapse was abetted by the USFS's difficulty in meeting politically imposed timber targets while trying to simultaneously comply with a veritable onslaught of environmental laws. Applicable laws, which interacted in unforeseen combinations and yielded unforeseen consequences, proved difficult if not impossible to follow while meeting mandated timber yield targets. In the late 1980s the cutting and concomitant fragmentation of what old-growth national forests remained in the Pacific Northwest was even more pronounced on the checkerboarded (alternate one-square-mile sections in public and private ownership) Oregon and California railroad lands managed by the BLM.

The environmental community—growing steadily and dramatically in numbers, monetary resources, political and legal acumen, and public empathy—was increasingly resistant to the continued cutting of remaining old growth to meet timber targets. Environmentalists became even more concerned as harvesting of remaining old growth was taking place at ever-higher elevations and on ever-steeper terrain.

When the northern spotted owl, a species dependent on old-growth forests, was listed as threatened, the USFS made several attempts to develop plans to conserve adequate old-growth forests so that it would be deemed unnecessary to list—as either threatened or endangered—potentially hundreds of other species dependent on old-growth forest habitat. It was feared, quite correctly, that a plan that would pass legal muster would entail significant reductions in the amount and rate of cutting of old-growth forests, dramatic reductions in timber industry profits and employment, and decreased revenues to federal and county treasuries. The various plans that surfaced differed only in the degree of their impacts and likelihoods of meeting the requirements of the ESA.

Even these early suggested solutions, minimal as they turned out to be, were not politically acceptable to the state governments of Oregon, Washington, and California and the congressional delegations of those states or to the administration of President George H. W. Bush. That produced a bubbling political witches' brew exacerbated by a startling level of intertwined political arrogance and ineptitude, which led inexorably to legal consequences.

Any solution to a public land management issue that survived scrutiny and challenge would inevitably be judged as politically possible, fiscally feasible, socially acceptable, and in compliance with applicable law. Politicians and political appointees would do their assigned jobs and then assure that their management was within the scope of laws and regulations. But when such assumptions are challenged in federal court, it is the federal judges who decide.

By late 1989, it finally became obligatory for the administration to come to grips with the deteriorating situation. The four heads of the relevant agencies—Dale Robertson, chief of the USFS; Cyrus Jamison, director of the BLM; David Ridenour, director of the National Park Service; and John Turner, director of the U.S. Fish and Wildlife Service—drafted me to recruit a team of experts to develop a scientifically credible strategy for the management of the northern spotted owl and its associated ecosystem. I was given carte blanche to select the team and determine the approach, and I was assured of whatever resources were required, granted freedom from interference, and given six whole months to do the job.

I had no personal experience with northern spotted owls. Further, I had no real firsthand understanding of the ecology of forests west of the Cascades. My job was to organize, lead, and direct the effort, not to be the lead technical expert. I recruited a core team with appropriate experience to formulate a viable approach and then jointly—along with me—shoulder responsibility.

The core team included wildlife scientists Drs. Eric Forsman, Barry Noon, and Jared Verner from the USFS's research division; Dr. E. Charles Meslow, director of the U.S. Fish and Wildlife Service's Cooperative Wildlife Research Unit at Oregon State University; and Joseph Lint, lead wildlife biologist from the BLM. Twelve other senior wildlife biologists and scientists from academia and state agencies in Washington, Oregon, and California were fully involved in the process. A number of other biologists from various federal and state natural resources agencies and universities provided support on an ad hoc basis.

The team had a jawbreaker title: the Interagency Scientific Committee to Address the Conservation of the Northern Spotted Owl—or ISC for short. During its first five months, the ISC reviewed pertinent literature, visited known habitats over the owl's complete range from northern California into Canada, and reviewed available data on owl locations and associated habitats. An initial management plan was based on what old-growth habitat remained, and a system of reserves of old-growth/late-successional forests on federal forest lands was the plan's backbone.

Reserves were selected based on five principles: (1) larger reserves are better than smaller; (2) many reserves are better than fewer; (3) reserves closer together are better than reserves farther apart; (4) intact reserves are better than fragmented reserves; and (5) reserves connected by suitable "dispersal" habitat are preferable to those not so connected.

The ISC and supporting staff were ensconced in Spartan quarters in an old office building in Portland, Oregon. Meetings and work sessions—except those during the preparation of the final report—were fully open to observers, although supervisory personnel from the federal agencies involved were excluded to preclude any impression that they influenced ongoing efforts.

Initially, observers were plentiful. Then their numbers dwindled steadily, reaching zero after several weeks. The hours were brutal and moments of excitement or controversy rare. For the most part, the work could be described as intense, tedious, difficult, boring, exhausting, and technically demanding.

What the ISC considered the appropriate number, size, and best arrangement of reserves was designated on maps as the foundation of the final strategy. The ISC report, to my consternation, was commonly and inappropriately referred to by the media and in politically oriented press releases as the "Jack Ward Thomas Report."

That was unfair and inaccurate, as it exaggerated my contributions, which were largely limited to organization and leadership, and simultaneously shorted the efforts of the scientists and biologists who had the pertinent knowledge and experience and did the real work. I didn't deserve, desire, or need the credit. I just wanted to complete my assignment and get back to La Grande, my wife and kids, and my own research.

After six long months of ten- to fourteen-hour days and six- or seven-day workweeks, the ISC completed its assigned tasks on time. It was clear that impacts on the timber industry and on county and federal treasuries would be sudden and dramatically negative. All concerned—the timber industry, state and county governments, federal land management agencies, environmentalists, elected officials at all levels, and the four agency heads—nervously awaited the ISC's report.

All concerned knew, or could and should have anticipated, that the ensuing political attacks would be immediate, intense, and negative and would come from all quarters. At best, the ISC hoped for a "thumbs up" from the scientific community, which, for the most part, was forthcoming. Significant social, economic, and political adjustments would be required in the status quo to implement the ISC strategy. There seemed to be no viable escape route for the administration from the ISC report, which, after all, *was its report.*

Any decision by newly elected President William Clinton was sure to draw attack—legal, political, and economic—from all directions. However, it was clear that the ISC strategy, or something very much like it, would be required to ensure compliance with the ESA, which would most certainly engender a dramatic political backlash.

Clearly, no one with a dog in the fight would be happy. The hard-core activists of the environmental persuasion accused the ISC of "selling out to the timber interests." Those concerned with the welfare of the timber industry regarded the ISC report as "overkill" and "collusion with environmental extremists." Elected officials tried to split the difference, posturing and pandering to whichever constituents they most valued and most needed for reelection. Our in-house predictions were borne out—across the board and in spades.

The ISC presented its report to the heads of the four federal natural resources agencies in a hotel near the Portland airport. When the ISC's briefing team arrived at the meeting site, several hundred pro-timber industry demonstrators replete with placards and banners were waiting. Clearly, the time, place, and purpose of the briefing had been leaked by a government source, but only to one side in the coming political melee.

The four agency heads listened intently to the ISC's presentation. John Turner, a wildlife biologist by training, congratulated us on the technical quality of our efforts. Dale Robertson was grim but professional and appreciative. Ridenour said very little, as the ISC report had no direct impact on the national parks. Then Jamison exclaimed, in the middle of my presentation, "This will cost me my political career!" The ISC was—to a man—stunned by his remark. Thousands of people would lose their jobs, lumber mills would close, rural areas would face severe economic dislocations, county tax bases would be diminished—and even then the northern spotted owl and the old-growth ecosystem would still be at risk. Yet Jamison seemed most concerned about his political career. We were informed that we were dismissed upon submission of our final report.

The ISC walked out of the front door and through the crowd of shouting, placard-waving demonstrators, many of whom, quite legitimately, seemed more frightened than angry. We tried to put ourselves in their shoes and knew that was what we would likely feel. What person so impacted would have felt differently? That made us even more aware of the significant consequences that would surely ensue if our recommendations were adopted. Now we had less than a week to meet our deadline for submission of our final report.

Shortly after the final draft report went out for peer review, "professional environmentalist" Andy Kerr, refusing to identify his source, announced to the press that the "environmental community" had acquired a draft copy of the ISC's report. All the peer reviewers, credentialed scientists with appropriate experience, had been monetarily compensated for their services and had signed agreements to maintain confidentiality, not to photocopy any of the materials, and to return all materials to the ISC. Clearly, at least one of that group had violated that agreement—and that trust.

We suspected the leak had come from a conservation biologist at the University of Chicago whose credentials included modeling work relative to the population dynamics of the northern spotted owl. When I confronted the biologist in a phone call, he feigned innocence at first, then admitted violating his contract. He seemed shocked and contrite—more, I thought, because he had been found out than for the lapse in integrity. He said an environmental lobbyist had wheedled the draft report out of him and shared it with Kerr, the lead spokesman for the environmental warriors involved in the ongoing political drama.

Oddly, I felt sorry for this reportedly brilliant young scientist; he was, to my mind, hugely naïve relative to political and ethical intrigue. In my opinion, his "friends"—much more experienced and skilled political operatives driven by the "purity" of their cause—used him carelessly and poorly. For some combatants, winning was everything. As one such protagonist advised me later, "Grow up! There ain't no rules in a knife fight."

The government's lawyers advised me to ignore the intrigue. No laws had been broken, and there was nothing to be gained in chastising the young scientist. Although he had violated his contract, he didn't offer to return the fees for his services and I didn't ask.

Contacts within the environmental community informed me that some environmental groups, now engaged in vigorously attacking the ISC report and its credibility, had initially been stunned by the magnitude of the proposed strategy. One of them said to me much later, "We would never in a hundred years have dreamed of asking for so much protection for the northern spotted owl and its old-growth habitat." That didn't stop him from immediately attacking the "inadequate protection" provided. Science, politics, and money made for strange bedfellows.

Unfortunately, the leak eroded the credibility of the ISC in the eyes of some in the administration and some members of Congress.

Then suddenly we were finished. The clicking of keyboards, whirring of copy machines, and jangling of telephones stilled. A delivery truck picked up several hundred copies of our final report, to be shipped off to the offices of the involved federal agencies in Washington, D.C., for distribution as they deemed fit.

I took my secretary, Nancy Delong, who was serving as our office manager, out to a nice lunch in a real restaurant for a change and then sent her on her way home to Pendleton, Oregon, with my deepest gratitude. I asked if she wanted to come to Washington for the grand finale. She passed, saying that she had been away from family and her real job far too much and for far too long. I smiled. Down deep, I wished I could just go home and pull the covers over my head.

I returned to my office to clean out my desk. On the way, I walked through the maze of desks, computers, telephones, chalkboards, flip charts, slide projectors, typewriters, copy machines, and other paraphernalia that now stood alone and what seemed to be almost ghostly silence. The place had been a beehive of constant activity for months. Now I was all alone in the place, and it was eerily quiet. My footsteps echoed. Sitting in my desk chair was a framed reproduction of a painting by Mary Englebreit—a parting gift from an anonymous colleague.

The picture was a stylized cartoon titled in red script across the top "Never Look Back." Depicted was a path that wandered through the woods and split up ahead at a fork in the trail. A small pilgrim, with a stick over his shoulder from which a bundle of possibilities dangled, resolutely strode toward the fork. Two signs were nailed to a post that marked the fork in the trail. One pointed to the right and read "The Rest of Your Life"; the other pointed to the left and read "No Longer an Option." A small owl with big, wide-open eyes—I saw it as a northern spotted owl—perched atop the pole.

I smiled, picked up the picture, held it at arm's length, and pondered the message. At first, I thought of the owl as a symbol of wisdom. I hoped that we had been wise in our efforts. Then an alternative interpretation came to mind. In some Native American cultures, owls are viewed as harbingers of evil and of death. My life, the lives of my family, the lives of close colleagues and their families, and the lives of all those impacted by our report were changed—inevitably and forever and none of them I suspected for the better.

The ISC had been sent on a journey, certainly not one of our choosing. Inevitably, we had come to one fork after another in our path to our final conclusions. In retrospect, we really had no choice about which forks to take. We just hoped we had. The decisions were essentially preordained. It was also a fork in the trail for the management of the public's lands—and not just in the Pacific Northwest.

What lay ahead down this new trail? Though I wasn't sure, I knew that we had, in the words of the first chief of the USFS Gifford Pinchot, "plowed new ground" and the management of federal lands would be forever changed. I hoped that was for the better.

I picked up my briefcase, tucked the picture under my arm, dropped my office keys on the desk, and left the building. I pulled the door shut behind me for the last time, listened to the lock snap shut, and hurried to catch the trolley to my motel. I didn't look back. Early tomorrow, I would be on a plane to Washington, D.C., for what I thought of as the grand finale. There was an emptiness in the pit of my stomach as I pondered the prophecy embedded in the painting.

Hearings before committees in the House and Senate followed. They were, to my disappointment and consternation, more about providing members of Congress platforms for "cover your ass" polemic castigations and political posturing than about imparting information or gaining understanding. The hearings were, for the most part, filled with "sound and fury signifying nothing." I remained calm throughout by telling myself that this was just my fifteen minutes of fame and that "this too will pass."

Four days later, I was in the air on my way home from Washington. When I arrived in Portland, I would transfer to a commuter plane to Pendleton. Margaret (if she was well enough in her battle with cancer) would pick me up for the fifty-mile drive over the Blue Mountains and, at long last, home.

By the time I made my way to the "bullpen" that served as the boarding point for all the commuter airlines out of Portland, I was physically and emotionally exhausted. I sprawled on a bench with my arms draped over the empty seats on both sides and stretched my legs out in the aisle. Several people seated just across from me looked me over as they whispered back and forth.

Finally, one leaned forward. "Say, mister, did anybody ever tell you that you look just like that spotted owl biologist guy—Jack Ward Thomas?"

"Well, yes, several people have told me that over the past few months."

He gave me a sympathetic look. "Geez, man! That must really piss you off!"

I nodded, "Sometimes."

As I write this, some two decades later, I can look up from my desk and study that picture of the little pilgrim approaching a fork in the trail. Over the decades since, I have taken even more forks in my trail—some significant and all unforeseen. But now that I am eighty years old and fighting cancer—and likely losing—the trail seems to have straightened out toward the horizon and there is not much further to go.

Now there is just one more significant fork in the trail ahead. I can't see it just yet. But no doubt it lies ahead, just over the rise. I am in no hurry to get there. But, once again, there is no choice. I wonder if my old friend, the northern spotted owl, will be sitting on the signpost.

"SHIT, KID, WE'VE ALREADY CUT THE OLD GROWTH!"

As required by the ESA, on June 22, 1990, the U.S. Fish and Wildlife Service declared the northern spotted owl to be threatened. Unfortunately, that inevitable decision had been delayed, for understandable political reasons, as long as possible. That delay rendered compliance all the more socially, economically, and politically painful.

The bill came due in June of 1991 when a federal judge, William Dwyer in Seattle, shut down all timber harvests from federal lands within the range of the northern spotted owl in Washington, Oregon, and northern California. The injunction was to remain in force until a "credible plan" had been adopted to protect the owl—and the ecosystem it represented—from extinction while putting both on a road to recovery. Then the legal and political noose tightened another notch when the U.S. Fish and Wildlife Service listed as "threatened" the marbled murrelet (a small diving bird that nested in old-growth forests) and several subspecies of salmon (some of whose best spawning habitat occurred in old-growth forests).

Those whose economic welfare depended on continued harvests of old-growth timber, along with their rightfully concerned elected officials, desperately wanted to believe that there had to be a better way—cheaper way—than that prescribed by the ISC. They demanded that other teams of technical experts be convened to come up with an alternative credible management plan that would allow continued cutting of significant amounts of timber from old-growth forests on federal lands.

The political and social backlash increased as each new suggested alternative—including delays and "more studies"—increasingly failed the smell test for credibility. The political noose grew tighter. Those delays and the associated publicity ultimately served the timber industry poorly, and the legal noose simply drew ever tighter, despite the understandable, if not fully justified, howls of politicians at local, state, and federal levels. Members of Congress had passed the operative laws, and they could have changed those laws. But they couldn't or wouldn't do either—and made no serious efforts to do so.

Instead they spent their time and energy posturing and casting stones at those who were struggling to comply with the myriad of laws that meshed ever more poorly. We learned that, with rare exceptions, elected officials served, or pretended to serve, the interests of their primary constituents. Once again, "money talked and B.S. walked."

Then in an unprecedented move the ISC was put "on trial" in an adversarial hearing in front of an administrative law judge in Portland. The "trial" was arranged by the secretary of the interior, Manuel Lujan. He thought he had uncovered a potential "escape clause" embedded in the ESA—the right to convene the Endangered Species Committee, a.k.a. "the God Squad." If this committee deemed the consequences of compliance with the law too onerous to bear, the God Squad could allow extinction or extirpation of a threatened or endangered species and, in theory at least, the stability of the associated ecosystems. The God Squad was composed almost entirely of political appointees of the president. Implied in the secretary's less than cleverly "leaked" cover story was the hope of exposing the technical ineptitude and/or malfeasance of the ISC.

The ISC's core team was put under oath and grilled by an "odd couple" of "prosecutors" appointed by the secretary of the interior: a lawyer employed by the BLM and a lawyer in the employ of the timber industry. The ISC team came to refer to them, facetiously, as the "dynamic duo" as they embarrassingly bumbled their way through the proceedings.

The day before the ISC team was to testify, Secretary Lujan's attorneys boasted to the press, "Tomorrow, we will defrock the high priests of the cult of biology!" The situation was bizarre—and becoming more so by the day. The Bush administration was putting its own employees—scientists and experts that it had assigned and charged with the task—through a hearing with the full hope and intent of shredding their credibility.

After my turn on the witness stand being "defrocked" for several hours, the hearing officer declared a recess. As I stepped outside for a breath of fresh air, I was approached by the grand old man of timber industry lobbyists in the Pacific Northwest. For many years he had been the industry's primary spokesman, advocate, and dispenser of support to favored political candidates. He was a tall, lean, imposing figure, though now a bit stooped by age, with bushy gray hair and eyebrows. He was, understandably enough, not pleased with the ongoing declines in timber yields from federal lands in the Pacific Northwest. He focused his ire on me as the nominal leader of the ISC.

He stood close and said, in his deep gravelly voice, "I know you boys think you've saved the old growth. Shit, kid, we've already cut the old growth! Live with it! What you've saved is nothing but the leftover crap. And in the end we'll get that, too!"

He had a point. Most of the best of the old-growth forests on highly productive low-elevation growing sites had already been felled. But the ISC didn't set out to "save the old growth." Our assignment was to comply with the requirements of the ESA as interpreted by the courts—nothing more and nothing less. And that is exactly what we did to the best of our ability.

As I sat on a bench at a bus stop in the canyon created by high-rise buildings and watched the traffic surge by, I thought about the remaining old-growth "crap" that now made up the heart of the reserve network created to accommodate the northern spotted owl and hundreds of other species. I thought, "Well, it may be 'crap,' but it's pretty damned magnificent crap."

I could not help noticing a sign carried by one of a number of demonstrators, both pro and con, outside the courthouse. Was the ancient prophesy from Isaiah 34:13 germane? "And thorns shall come up in her palaces, nettles and brambles in the fortresses thereof, and it shall be a habitation of dragons, and a court for owls."

In the end, the God Squad, which included Secretary Lujan, who ordered the inquisition, considered the transcripts of the proceedings and in nearly all aspects supported the ISC strategy. The "trial," or "defrocking," was a humiliating misstep—technically, politically, and legally—for the George H. W. Bush administration. Whether it was a wise political decision forced upon Secretary Lujan was not for us to judge, but it left a bitter taste.

The ISC team members—especially Drs. Noon, Verner, Meslow, and Forsman—were simply too formidable as witnesses for the dynamic duo of attack dog lawyers. The ISC did have an advantage in that our assignment was neither difficult nor complex—we simply had to tell the truth, the whole truth, and nothing but the truth. Of course, it didn't hurt that the ISC's attorney, Patrick Parenteau from the USF&WS, had run circles—tight circles—around the dynamic duo. He simply had the dynamic duo outgunned.

As a sop, the God Squad allowed the BLM to proceed with several low-impact old-growth timber sales before coming into line with the ISC's management strategy. Now the status quo would remain in place, at least until after the pending presidential election between President George H. W. Bush and Governor Bill Clinton. Issues over the northern spotted owl, marbled murrelet, salmon, and old-growth timber would play a role in that election, especially in Oregon, Washington, and California.

Ironically, the few old-growth timber sales on BLM lands approved by the God Squad were never cut—more salt rubbed into Secretary Lujan's and BLM director Cy Jamison's self-inflicted wounds. "His bread I eat, his song I sing."

The God's Squad's decision was a hugely significant one for the management of the public's forest lands in the Pacific Northwest. It would take a couple of years and considerably more effort for the administration of newly elected President Clinton to bring the ongoing fiasco to a conclusion that has lasted, in spite of all attacks, to the present time. And "the beat went on and the beat went on."

NO DEATH THREATS AT HOME

After the ISC report was released in April of 1990, newspapers and television newscasters reported widely varying estimates of the plan's impact on the timber industry, employment, and the economy of the Pacific Northwest. The estimates of jobs lost ranged from less than 10,000 (from the hired guns on the environmental side) to over 150,000 (from the hired guns on the timber industry side). Timber industry representatives and the labor unions protested that the plan went much too far in setting aside old-growth forest habitats—and threatened political action. Environmental groups insisted that it didn't go nearly far enough—and threatened legal action. Both sides scapegoated the ISC team.

The ISC members resumed, as best we could, our regular jobs and lives. USFS top-level administrators, concerned about the uproar and especially the identification of the report as the "Jack Ward Thomas Report" and the ensuing threats, offered to sequester me and my family at an out-of-the-way resort and put a guard on our house in La Grande for a couple of weeks.

My immediate boss, Dr. Charles "Charlie" Philpot, director of the Pacific Northwest Forest and Range Experiment Station, was likewise concerned with my welfare and safety and my family's as well. As the USFS had research going on in Guam, Philpot graciously offered me a research position in Guam, reasoning that out of sight was out of mind. He promised that, when the smoke cleared, I could return to my job in La Grande. It was a generous offer, one that Margaret and I genuinely appreciated. At another time and under different circumstances, Guam might have been an exciting and interesting assignment for a few years. But we opted for being sequestered for two weeks.

After Margaret and I returned to La Grande, we opened each issue of our regional newspaper, the *Oregonian*, with a twinge of trepidation. For almost a month, nearly every edition contained at least one news story, often on the front page, which addressed the validity and the probable social and economic consequences of the adoption of the "Jack Ward Thomas Report."

Our family had lived in La Grande for nearly two decades. Margaret and I were active in the community and acquainted with hundreds of people. Margaret was perhaps better known locally as the result of her musical endeavors, church activities, and club work than I. Most folks, and nearly all of our friends, treated us as they always had. A few, however, ceased to speak to us or otherwise made their negative feelings toward us obvious. That was painful. On the other hand, we could not help but empathize with their distress.

One morning, as I went to get into my Ford pickup to drive to work, I noticed that the hood was not fully closed. I retrieved a flashlight from the truck's gear box and peered into the engine compartment without opening the hood. There, plain as day, were three sticks of dynamite!

After my heart restarted, I took a closer look. The "dynamite" turned out to be road flares. But I understood the message the flares were meant to convey. The road flares *could* have been dynamite. I *could* have turned the key and blown the truck, me, and maybe my house and family sky high.

The message was clear, I should be *afraid—very afraid!* I called the police. They found nothing they could follow up on. But, bless their hearts, they did what they could for our protection, driving by the house a few times each day and night. We were grateful.

Several weeks later, as I walked out the front door to go to work, I knocked over a Molotov cocktail that had been leaning against the screen door. Again the point was crystal clear. "Please note the Molotov cocktail. We could have lighted several such devices and thrown them against both doors and through a couple of windows of your house. Your family would have had a hard time getting out alive. *Be afraid! Be very afraid!*"

As much as I hated to admit it, these threats were getting to me, especially now that they extended to my wife and boys. Maybe we should take the assignment in Guam? Margaret was simultaneously gentle and whet-leather tough. There was no way, at least not yet, that she was going to be scared out of her home. She was slowly dying and standing tall.

She reasoned that there was no real intent to kill me—or her and the boys. After all, anyone with a hunting rifle could easily kill me when I walked out the door some morning to go to work or ambush me while I worked in the woods. If they wanted to blow up the house, they could do that easily enough. But she agreed that it might be well for the boys to stay with friends for the next few days or weeks.

A week later, on a Saturday night just after midnight, the telephone beside our bed rang.

The caller asked, "Jack Ward Thomas?"

I sat up on the edge of the bed, "Yes?"

The caller slurred a word here and there. "I and a bunch of other hard-working honest Americans are losing our jobs because of you, you rotten communist bastard!"

Now I was wide awake and, to my surprise, ice-cold calm. I thought, never let the bastards make you sweat. "I can understand that, sir. Is there anything else?"

I listened to more heavy breathing. "You commie sonofabitch, you best be looking over your shoulder. All the goddamn time!"

Oddly, I wasn't scared and I wasn't even angry. I was just simply worn out—nagging fear is a wearisome thing. Then my admittedly perverse sense of humor took over. "Sir, I must ask if this is a bona fide threat?"

I could hear heavy breathing as my caller pondered his answer. "Damned right, this is a threat, you stupid sonofabitch!"

I replied in my most officious tone. "Sir, the USFS manual clearly outlines official policy for dealing with death threats. I cannot—repeat, *cannot*—accept death threats at home. Death threats can only be accepted at the appropriate USFS office between the hours of 8:00 A.M. and 5:00 P.M., Monday through Friday. The number is xxx-xxx-xxxx. Inform the receptionist that you wish to issue a death threat. The receptionist will, at the beep, record your threat and see that I get it. Again, that number is xxx-xxx-xxxx."

I pictured the caller writing down the phone number.

I said, "Have a pleasant evening," and hung up. He didn't call back that night—or ever. No death threats were received at the office. I was almost disappointed.

LUNCH IN FORKS, WASHINGTON

After the release of the ISC's report, the attention of the press in the Pacific Northwest, and to a lesser extent the national press, was focused on the dramatic negative social and economic impacts of the ISC plan. Understandably, the media, the gamut of interest groups, and politicians zeroed in on the probable impacts on people, local governments, and the economy. It was clear that the adverse impacts would be significant and concentrated in rural communities most dependent on the timber industry. And believe me, money talks!

Estimates of job losses were frightening enough, but the estimates lacked the context of human faces. Did the hundreds, even thousands, of demonstrators who were rallying and protesting out of fear, anger, and despair really believe they would, or even could, change the inexorable course of events without changing laws? Likely not, but they had to try, as I might have likely done in their shoes.

The Endangered Species Act was now recognized as the nine-hundred-pound gorilla of the plethora of environmental laws that had been so casually passed by Congress in the 1960s and 1970s—and signed into law by presidents of both political parties. In fact, the ESA had passed without a single dissenting vote in Congress. The long-delayed consequences from the interactions of adherence to myriad longstanding directives to federal land managers, newer environmental laws, and subsequent court decisions had finally come to a head in the early 1990s.

After the ISC strategy was released and rejected by President George H. W. Bush, Congressman Eligio "Kika" de la Garza of Texas, chairman of the House Agriculture Committee, appointed a working group to develop, this time around, an *array* of alternatives for the conservation of the northern spotted owl that included estimates of related impacts on jobs and the economy. It was assumed that Congress would enact one of the alternatives into law. Likely, it was de la Garza's chief of staff for the House Agriculture Committee, James R. "Jim" Lyons, who came up with the plan.

The working group included forestry policy expert Dr. John Gordon of Yale, forest economist Dr. K. Norman Johnson of Oregon State University, forest ecologist Dr. Jerry Franklin of the University of Washington, and me. A timber industry spokesman tagged the team the "Gang of Four," using the pejorative term that had been given to a group of politicians tried for corruption in Communist China. The Gang of Four delivered, on time, an array of options, including estimates of impacts on jobs and the social and economic status of the region. Chairman de la Garza quickly discerned that he had given birth to a truly hot political potato that showed no signs of cooling off. He postponed any action by the House Agriculture Committee until after the upcoming presidential election.

In the meantime, environmental activists sued the federal government for noncompliance with the ESA in the case of the northern spotted owl—and won. All timber harvests from federal lands, both USFS and BLM, within the owl's range ceased. As this was going down, my name and sometimes my picture began to appear routinely in newspapers and on TV as the "northern spotted owl guy."

The town of Forks is located on the Olympic Peninsula in Washington State. The ISC had recommended putting nearly all the remaining old-growth forests on federal lands there off limits to logging. Most of the folks in Forks, directly or indirectly, made their living from the cutting and milling of old-growth timber cut primarily from public lands. Since the late 1940s, the timber industry had been an economic "sure thing" for the town of Forks—the only "sure thing." But now that certainty had quite suddenly morphed into a social, economic, and political disaster. The people of Forks were increasingly hurt, angry, and scared—a potentially volatile combination—as their world unraveled ever faster.

This cartoon, personally inscribed by the artist for Jack, appeared in The Oregonian, *May 30, 1990.*

Shortly after Judge Dwyer's ruling, I was driving through Forks in an apple-green USFS pickup with identifying decals on the doors when I encountered a human dummy hanging from a lamppost. There was a sign around its neck—"Jack Ward Thomas." A bit farther down the road, several effigies of owls hung from lampposts. Being both hungry and in need of a caffeine fix, I stopped at a restaurant.

It was early afternoon and the place was not crowded. I sat down in a booth and took off my baseball cap and sunglasses and identified several loggers seated at the bar by their weathered faces, hickory-striped shirts, suspenders, staged-off pants, White's boots, a few baseball caps bearing the logo of the Stihl chainsaw company, and a couple of battered hard hats.

The most senior looked at me over his shoulder. He turned slowly around on his barstool, stood, and walked over. He was a big man—a hard six feet plus a couple of inches and weighed around 240 very solid pounds. He came straight to the point. "Are you Jack Ward Thomas?"

I thought, "Oh shit!" and took a deep breath. It seemed prudent to stand up while keeping my hands out of my pockets in case there was a punch on the way. "Yes, sir." I extended my hand. "And you are . . . ?"

After a long moment, he shook my hand. His was a big hand, hard and rough with calluses. He told me his name and asked, "Can I talk to you for a minute?"

"Sure thing, please join me." I signaled to the waiter for another cup of coffee and whatever my new acquaintance was having.

For a moment, my companion sat quietly studying his interlaced fingers on the table. Then he turned and looked me in the eye. "Do you have any idea what you are doing to me, my family, to this community? Our whole goddamn way of life is unraveling. Worse yet, there doesn't seem to be anything we can do about it. We don't even know who can maybe change things. Nobody ever talked to us about what was going down. Not one solitary soul. Do you folks understand all that?"

He answered his own question. "I can't believe you do. If you knew, if you really understood, I can't believe you would do what you're doing. No damn way."

In a low voice, I gave my new friend the short version of the requirements of the myriad individual laws, interactions between the laws, judicial decisions, and biology of the northern spotted owl and the ecosystems that made up its habitat. His eyes figuratively glazed over—it was too much information.

In the process, it came to me that the developing situation more and more resembled a classic Greek tragedy—the circumstances could only lead to an inevitable tragic conclusion for him and his fellows. Only the "playwrights"—the Congress and the president— could alter the script. And they, collectively and individually, were diving under the tables and covering their collective asses.

Otherwise, the developing interaction of federal laws and the tattoo of decisions of the federal courts would become ever more constraining on forest management activities on federal lands. The actors in the drama—loggers, mill workers, government employees, lawyers, judges, environmentalists, stockholders, scientists, politicians— could, in the final analysis, change nothing so far as the inevitable outcome was concerned.

We talked for over an hour, enough time for us to begin to see each other as human beings instead of stereotypes. He was just over sixty years old and had dropped out of high school at sixteen to work in the woods. In those days, a hard-working skookum young man skilled in the logging trade could step directly into the middle class. He married a year after he went to work in the woods and within another year was the father of the first of four kids. By his fortieth birthday, he had saved enough money, had good enough credit, and acquired enough skills in logging and in handling men and equipment to put together his own logging outfit. At the time, his chosen profession was an honored, even heroic profession. The mystique of Paul Bunyan and Babe the Blue Ox still lived on the Olympic Peninsula. He was a successful small businessman and justifiably proud of himself and his way of life.

Now it was all falling apart around him fast, including his company, profession, future, and (even more important to him) the future of his sons and sons-in-law and their families—all in the timber industry. Now suddenly he was "upside down" on the loans for his logging equipment, which, no matter how new or how well maintained, was suddenly glutting the market.

One of his sons had been killed in Vietnam. He and his wife were helping bring up two grandchildren. He looked down at his clasped hands. "Just what the hell am I supposed to do? I read in the paper where the government says, 'Not to worry, retraining will be available—in computers maybe.' Hell, man, I'm over sixty years old. I'm smart enough, but my smarts are woods smarts and people smarts and business smarts, not the kind of smarts you get out of books. And now somebody is going to 'retrain' me? To do what? What will I be earning? Minimum wage or, if I'm lucky, a little more?"

He held up his hands, palms toward me with fingers spread. The fingers were thick and muscular. The calluses on his palms seemed to form a work glove of sorts. The tip of one finger was missing. "Do these look like hands that would fit a computer keyboard? These bright ideas about 'retraining' folks like me must have been thought up by some dudes in three-piece suits over martinis in some fancy bar in Washington, D.C."

He put his face in his hands and, for the briefest of moments, choked back emotion. I looked away, not daring to say anything—not a word. I, too, was fighting back tears. What could I say? I was just as unable to rewrite or redirect the unfolding economic, social, and environmental tragedy as he.

Suddenly he shoved back his chair, stood, and offered me his hand. I stood and took it. Over my protest, he picked up the check and went back to his friends. As I walked out into the bright sunlight, I looked back into the shadows and waved good-bye with a slowly executed military-style salute. He returned the honor. His companions just looked at me—without expression.

I drove away. Several miles out of Forks, emotions welled up that I could not choke down, and I pulled over and let the tears come. People, real people—with hopes and dreams and families—were paying a terrible price for what, as ordained by law and court decisions and belated concern for ecosystems thousands of years in their creation, now had to be faced.

The steadily increasing timber harvest from public lands had evolved from being considered "good forest management" to "sins of the past" to a new and belated paradigm of "ecosystem management." It didn't matter how right and how logical those "sins" had seemed when committed.

Times and circumstances change. The bill, long delayed and even denied, had suddenly come due. My colleagues and I were simply the delivery boys. Being on the receiving end, no doubt, was dramatically worse.

I, too, was exhausted and sick at heart with the whole state of affairs, but I still had a job. Oddly, I felt guilty about that. Maybe that didn't make any sense, but that was what I felt. Maybe that was what a military head shrinker might label "survivor's guilt."

Each member of our working group of scientists, in their individual ways, felt the pressure and anguish of those who would lose their businesses and jobs if one of our suggested alternatives were adopted. The impacts had come home to me. The exchange with my new—dare I think—"friend" taught me more than attacks from the administration, hostile hearings in Congress, parades of chanting sign-carrying protesters, and anonymous threats. My new friend, and his and his family's plight, was to come to me again and again in my dreams.

I was reminded of a quote by President James Madison that seemed especially germane:

> It will be of little avail to the people, that the laws are made by men of their own choice, if the laws be so voluminous that they cannot be read, or so incoherent that they cannot be understood: if they be repealed or revised before they are promulgated, or undergo such incessant changes, that no man who knows what the law is to-day, can guess what it will be to-morrow.

TALK TO HIM IN PRINCETON

After the release of the ISC report in April of 1990, pressure was building for a more "politically acceptable" course of management of old-growth forests with less negative impact on the timber industry, but one that would have some chance of standing up in federal court.

Mark Rey, then the point man for the American Forest Products Association, put together a group of well-qualified wildlife and respected ecologists in the direct or indirect employ of major timber companies. Joe Lint, lead wildlife biologist for the BLM and a former member of the ISC, was included. They were tasked with developing an alternative strategy to assure the welfare of the northern spotted owl while dramatically lessening the impact on timber yields.

Based on my conversations with several of the players, I thought that most believed they were on a "fool's errand," but I wished them well. Maybe—not likely, but just maybe—they might uncover approaches that we had failed to consider.

Mark Rey was in his mid-forties and the proud possessor of a razor-sharp mind. He was of medium height with coal black eyes and hair and a dark complexion. He sported a mustache and grizzled goatee. I had never seen him when he wasn't wearing a dark suit and tie. He spoke in a low voice in carefully cultivated and measured cadences that demanded attention. Some of us, even to his face, referred to him as "Darth Vader." In retrospect, it was a bum rap.

The question was, could the Republicans reset the clock relative to the situation in the Pacific Northwest? After all, they had taken over the Senate in the last election, and some felt they had the power, and maybe the mandate, to try. If anybody could guide them to success through the maze to achieve their ends, it would, in my opinion, be Mark Rey.

Another potential last-gasp effort to avoid what seemed like the inevitable surfaced. Why not establish state-of-the-art "owl farms?" Maybe such farms could eventually produce hundreds of northern spotted owls for release into the wild—when and where needed. That raised a politically if not biologically germane question. How could a species be considered threatened or endangered if hundreds were alive in captivity and their offspring were being systematically being released into the wild?

A congressional committee announced a hearing to explore constructing and staffing facilities to hold captured spotted owls and raise their offspring to maturity. Such facilities could be replicated in whatever numbers and locations deemed necessary to fore-stall extinction. Money, apparently, was no object. It was a ploy, though not a very good one, to "keep hope alive." It likely would not fly with ecologists or with the federal judges, but it just might be a political winner.

When I received a telephone call describing the purpose of the hearing and "requesting" my presence, I thought it was a joke. I was wrong. I was to be joined by two key members of the ISC: Dr. Charles Meslow of the USF&WS and Dr. Barry Noon of the USFS.

We sat in front of the committee and listened to a presentation about meeting the requirements of the ESA by rearing spotted owls in captivity and releasing them into the wild. At the conclusion, the committee chairman leaned forward.

"Well, what do you think?"

Despite what I was thinking and feeling, I tried to be polite—always a good idea when you can be held in contempt of Congress. I commended the "innovative thinking" involved in the proposal; then I dropped the bomb.

"However, the suggestion is fatally flawed. First, we have no experience or knowledge that ensures that northern spotted owls can be successfully and consistently reared in captivity. Second, we have no way to reliably estimate the survival rates young owls after release into the wild. More significantly, such an approach would not be in keeping with the intent of the Endangered Species Act. Therefore, in our collective opinion, such a scheme would neither survive technical scrutiny nor pass muster with the courts. I direct your attention to the stated purpose of the act."

I made a show of flopping down on the table a big, fat, green-covered tome titled *The Principal Laws Relating to Forest Service Activities*. I read from section 2 of the 1973 Endangered Species Act: "The purposes of this Act are *to provide a means whereby the ecosystems upon which endangered species and threatened species depend may be conserved* [and] *to provide a program for the conservation of such endangered species and threatened species . . . .*"

That, in my opinion, should have put an end to the farce. But one congressman, who clearly had assigned lines to deliver in the ongoing drama, reminded us that money was no object. "How," he asked, "could anyone even pretend that a species was threatened or endangered if hundreds, maybe even thousands, of spotted owls were being raised and protected in captivity—and being routinely released into the wild?"

Dr. Meslow leaned forward and pulled a microphone close. He said something like: "Gentlemen, let me reiterate, in slightly different words, what Dr. Thomas told you. Rearing threatened or endangered species in captivity is, at best, a last-ditch effort to forestall extinction. Such an approach will not satisfy the requirements of the act. The 'conservation of ecosystems' is the underlying purpose and foundation of the act."

Some committee members, through body language and facial expressions, made it clear that they did not understand. We thought it more likely that they didn't want to understand.

Then a congressman noted that all of Meslow's and my degrees were from public land grant universities (i.e., state schools rather than elite private institutions). Weren't our degrees, he asked, in wildlife biology or forestry as opposed to ecology or conservation biology?

Fortunately, Dr. Barry Noon, who had yet to speak, held degrees from Ivy League schools, including a doctorate from Princeton. It was time to roll out the heavy artillery. I whispered jokingly to Barry, "Talk to him in Princeton."

Dr. Noon scooted his chair up to the witness table, moved a microphone into position, tapped it to make sure it was on, cleared his throat, and looked intently at the chairman. He then said, slowly and distinctly, in effect: "Mr. Chairman, we cannot proceed with what you suggest because it doesn't make a lick of sense."

Now, by golly, there was an explanation, enunciated clearly in Ivy League language, that anybody could understand! Everybody did understand what he said and meant. The meeting ended shortly thereafter. That diversion from reality did not come up again. The three Musketeers (or maybe the Three Stooges, depending upon one's point of view)—Meslow, Noon, and Thomas—dodged the reporters waiting to ambush us in the hall outside the hearing room. We were off for the airport and home. Enough was enough.

As we settled back in the cab, Meslow and I conceded that it was likely Noon's clarity of expression and flawless enunciation that turned the tide. Barry endured our praise for a few minutes until, smiling modestly, he indicated with an emphatic hand gesture that he had endured enough of our joshing.

On the way to the airport we pondered: just when would the politicians bite the bullet and allow adoption of a plan for managing the habitat of northern spotted owls—really the old-growth eco-system—that would stand up to court challenge? Or, alternatively, when would Congress summon up the courage to alter or clarify the laws, especially the ESA? We concluded that the deadlock and shutdown would continue at least until after the upcoming presidential election. And, sure enough, we were right about that—and the presidential election after that, and the one after that, and the one after that.

Sometimes forks in the trail don't lead anywhere.

"I THINK THIS WAS MADE BY GOD"

During the 1992 presidential campaign Republican president George H. W. Bush and third-party candidate H. Ross Perot vowed to revamp or revoke the ESA. Democratic candidate Governor Bill Clinton of Arkansas skillfully skated around the issue by promising to convene a "Forest Summit" in the Pacific Northwest shortly after his election. The purpose of the summit would be to come to grips, fully and finally, with the issue of the northern spotted owl, the marbled murrelet, salmon, and the old-growth ecosystems on public lands in the Pacific Northwest. Mr. Clinton won the election (Oregon, Washington, and California were critical to that victory) and, as promised, convened the promised Forest Summit in Portland in April of 1993.

Several members of the new cabinet were present, including Secretary of Agriculture Mike Espy, a former member of Congress from Mississippi. While the USFS was the largest agency in the USDA in terms of employees, federal land management is not something most agriculture secretaries know much or care about. Secretary Espy was aware that, for reasons of credibility, he needed exposure to and increased knowledge of forestry, grazing, and wildlife issues. It must have seemed like a good idea to him to see an old-growth forest—and maybe even a spotted owl.

Jim Lyons, who would soon be the undersecretary of agriculture with responsibility for the USFS in the new Clinton Administration, organized the summit. He arranged for Dr. Jerry Franklin of the University of Washington, the foremost expert on the ecology of old-growth forests of the Pacific Northwest, and me, chairman of the ISC, to conduct a tour of an old-growth forest stand for Secretary Espy. Requests from environmental and timber industry groups to tag along were denied.

On the drive from downtown Portland to a nearby stand of old growth—in a rented stretch limousine—Dr. Franklin gave Secretary Espy a short course on the ecology of old-growth forests. Kathleen "Katie" McGinty, recently appointed director of the Council on Environmental Quality, joined us at the site.

The four of us walked through the towering trees and the recumbent, slowly decaying skeletons of giant Douglas firs, stretched out in the shadows on carpets of mosses and ferns. The road noise dulled to a whisper. The air was still, cool, and moist. Rays from the midday sun penetrated here and there like searchlights into the twilight created by multiple layers of the canopies of the giant trees.

Dr. Franklin explained the intricacies of the old-growth ecosystem. Secretary Espy listened intently, saying little as his Old-Growth Forest 101 course progressed. At the appointed time, we arrived back where we started. Traffic sounds and the clatter of voices broke the spell of silence.

As we stepped into the sunlight, eyes blinking in adjustment, the members of the press pounced; overlapping shouted questions filled the air. A security detail shielded Secretary Espy from the pressing crowd. Just before he stepped into the limousine, he turned to face the press. Only one question was being asked, over and over and in a half-dozen ways.

"Mr. Secretary, what do you think of what you just saw?"

Secretary Espy was initially taken aback by the ambush. His reply came from the gut. "This was made by God, and I don't think we should cut it." And with that honest statement—political correctness aside—he stepped into the limousine and was whisked away.

Speaking as scientists, Franklin and I had talked with Secretary Espy about old-growth ecosystems and how they developed and functioned. We discussed ecosystem management and the ramifications of complying with the requirements of the ESA. But Secretary Espy had spoken from the heart.

FRIENDS YOU HAVEN'T MET

When I arrived in Missoula in 1997 as the new Boone and Crockett Club endowed professor in the College of Forestry and Conservation at the University of Montana, a local newspaper reporter asked me about the organizations I had worked for and the places I had lived. I counted the locations: Sonora and Llano, Texas; Morgantown, West Virginia; Amherst, Massachusetts; La Grande, Oregon; Washington, D.C.; and now Missoula, Montana.

She noted my Texican drawl and my dress—Levi's, khaki shirt, and beat-up White's smoke-jumper boots. She commented, "You must have really hated some of those places!" It was my ball—she had handed me a chance to play "good ole boy."

There wasn't one of those places, or jobs, that I didn't enjoy, admittedly some a bit more than others. Why? I loved my work and enjoyed being exposed to and learning about different ecosystems and human cultures. "Different" can be scary and debilitating or fascinating and stimulating. Margaret always saw a new duty station, a new job, and a new culture as fascinating and stimulating. Her attitude was infectious.

Margaret was outgoing while I was much more reserved. The personality tests I had been given at various USFS leadership training sessions over nearly three decades of service had consistently identified me as an introvert. She was enthusiastic. I was mellow. She was not afraid to appear naïve; I was. She joyfully joined in; I hung back. Over time she brought me around to conceding that life was a great, wonderful, and all-too-brief adventure to be fully embraced. I came to regret that it took her so long to break me to lead.

Margaret grew up in the "oil patch" in the desert around Monahans in West Texas. We met on a blind date in Fort Worth in 1955, when we were both sophomores in college. She was majoring in music education. I thought, from the moment I saw her, that she was the most beautiful girl I had ever met: five feet seven inches tall, slender, with blond-brown hair, brown eyes, a beautiful soft voice, and an appropriate West Texas accent. When the weekend was over, I told my mother that I had met the girl I would marry. It took time—and some skilled, serious courting—for me to woo her to my way of thinking.

In June of 1957 we married, a month after my graduation. She had graduated a year earlier and was teaching music in Edna in South Texas. For the first decade of our marriage we lived, very modestly, in the small isolated towns of Sonora and Llano in the Texas Hill Country. We had been married a decade when I took a job with the USFS in Morgantown, West Virginia—a major fork in our trail. Life was getting bigger, better, more vivid, more interesting, and more educational with each passing year.

Our Texican accents drew attention in our new location. She turned a potential handicap to advantage as what she called an "ice-breaker." When we lived in West Virginia our accents weren't too terribly much out of kilter. But when we arrived in Massachusetts, our drawls stood out! Her accent was even more pronounced than mine. She could somehow tell when somebody was about to comment on her accent. Then she would put her hand on their arm and declare, exaggerating her accent, "Why, y'all surely have a lovely accent. I just love it!" Silence would reign, and then almost inevitably the person who had those words taken right out of their mouth would break into laughter.

At every new duty station, Margaret—the icebreaker—would go into action. Soon my coworkers, their families, and our neighbors were comfortable around our table inside or in the backyard. In every new location, she saw to it that we became active in the Methodist church. In each place she sang in the choir, sang solos, and in some cases directed the choir. Her frequent parties for my professional associates and their families were a routine part of our lives and added to my recognition as a leader. She was, indeed, that rare "hostess with the mostest." In some of the places we lived, she taught music in the public schools, and she always taught private piano and voice lessons at home. Before I knew it, largely due to her efforts and personality, we would be a significant part of the communities in which we lived.

Margaret spotted folks I worked with who didn't quite "fit in" and gave them special attention. "Newbies" were always welcomed and put off to a good start. She was never on the TGD or USFS payrolls, but she surely should have been. We were, more due to her efforts than mine, something of a "dynamic duo."

Once, while we lived in La Grande, she organized a going-away party for a young biologist who had begun his USFS career as a "temporary employee." After several years of sterling uncomplaining service for a pittance, wildlife technician Scott Feltis moved up in the USFS hierarchy when he landed a job as a full-fledged wildlife biologist on a national forest in Wyoming. Margaret noticed that Scott's wife, Kath Ann, was visibly upset about the pending move and took her aside.

Kath Ann broke into tears. "Meg, I don't want to go to Wyoming. I don't want to leave La Grande. I won't have a single friend there. But I know I have to go. I can't let Scott down."

Margaret hugged Kath Ann close and held her for a few moments. Then she said, "Kath Ann, honey, y'all have dozens of friends in Wyoming—simply dozens! And some of them are really good friends that will last a lifetime. You just haven't met 'em yet."

Kath Ann wiped away her tears. She wasn't completely buying Margaret's story, but she did seem to feel a bit better. Things worked out just fine for Scott and Kath Ann. They indeed made dozens of friends who were waiting in Wyoming. Sometimes forks in the trail—even when they pay off in the end—are hard to take.

CHAPTER 5

A CHANCE TO MAKE A DIFFERENCE:
WASHINGTON, D.C., 1993–1996

In 1993, I became the thirteenth chief of the USFS—the only chief ever promoted to that position directly from a field assignment and only the second to come out of the research division. I expected to be in the chief's job for only a few years—I had already accumulated twenty-seven years of federal service when I was appointed and was eligible for full retirement benefits upon completing thirty years of service. And, sure enough, three years it was to be.

IT'S A LONG WAY

I was approached about becoming the thirteenth chief of the USFS—the so-called "Long Green Line"—in the early fall of 1993. The previous year and a half had been hugely stressful. I was dealing with the ongoing crisis over the spotted owl and old-growth forests. My sweetheart of nearly forty years was battling colon cancer, which had spread to her liver. The prognosis was grim: death in the short term without chemotherapy, or death after some longer term with chemotherapy. But in either case it was death in the end, and that fork lay not far down the trail. We made a pact to live as normally as we could for as long as we could.

In love with life and her family, Margaret chose chemotherapy, which meant periods of remission in exchange for bouts of nausea and hair loss brought about by a new round of chemotherapy. Oddly, the hair loss bothered her more than the bouts of nausea. I was mightily impressed with her intestinal fortitude and will to live. I didn't believe that I could have endured more than one round of such treatment before opting out. But she was always the tough one of our odd couple.

For a time, she bounced back from each round of chemotherapy and relished every minute of the "in betweens," as she called the periods of remission. As she rebounded after a round of chemotherapy, her hair grew back—now dark brown and without a trace of grey. As quickly as she could, she immediately resumed her life—our life—as if nothing were amiss. She lived, I think now, in a state of grace focused in the moment.

When newly elected President Bill Clinton convened the promised Forest Summit in April of 1993, I was invited to make a presentation. Before it was over, I was summoned to a meeting with President Clinton and Vice President Al Gore and asked to try again to address the forest management issue.

To the extent that I had any choice, I agreed under the following conditions: (1) I would lay out the approach; (2) I would recruit a team of scientists and technicians of my choosing, from both inside and outside government; (3) adequate resources would be made available; (4) this time around, the team would prepare an array of options from which the president could chose; (5) there would be no political interference and no political observers; and (6) the president would not expect to hear from us until we presented our final report.

President Clinton agreed, and I went home and went to work putting together a team of the best and brightest professionals. My liaison with the president was Jim Lyons, the new undersecretary of agriculture who would oversee the USFS and the Soil Conservation Service.

This effort was christened the Forest Ecosystem Management Assessment Team (FEMAT). Most of the players had been involved with one or more of the previous three efforts to address the crisis. I had never seen government activity, outside of war, move so rapidly. Within two weeks we were at work. Over the course of the next two months, over 200 folks—scientists, technicians, and support staff—were involved. Personnel came from federal agencies (primarily the USFS, BLM, and USF&WS), academia, state forestry and wildlife agencies, the timber industry, and environmental groups. Operations continued ten to fourteen hours a day for weeks at a time, with but a few days off. It was the most intense and challenging experience of my forty-year professional career. Three months passed quickly, and we were done—on schedule.

Margaret, ill as she was, insisted throughout that we "do *our* duty and stay the course." She was more and more thinking of the two of us as one entity. There was much at stake, both for the environment and for those who earned their daily bread in the timber industry. In her opinion, I was the best person to recruit and facilitate the activities of the best highly qualified, self-confident, individualistic ecologists and forestry professionals available to get the job done.

That group would have to work collaboratively under extreme pressure, with impossible timelines, to produce products that would stand up to intense scrutiny—technical and legal. With so much at stake, intense scrutiny was to be expected. When I suffered periodic spells of "weak-knees syndrome," she exhibited more confidence in me than I felt, but this was nothing new.

We lived in temporary quarters in a small apartment in Portland. Fortunately, that put her within easy reach of the specialized oncologists who were guiding her cancer treatment. But we missed our home, friends, and church in eastern Oregon. The combined pressures of her worsening illness and my work assignment grew steadily.

After my professional chores were complete and the products were delivered to the president, we went back to our home in La Grande at long last. By now, Margaret's condition had become hopeless, and we both knew it.

She never once complained. Aided and abetted by the USFS, I set up an office in our home so that I could work and be close to Margaret twenty-four hours a day when times were especially tough. As she wanted it, I served as her nurse and cared for her as best I could. The chemotherapy treatments were over now; increasing doses of pain-killers inexorably took their place.

But to our amazement, once the chemotherapy was over, she rallied and even seemed to recuperate. She pretended to believe that as long as her hair was growing, things were going well. After she recovered to the point where she could take care of herself and be out and about, I returned to doing some field work. That was a blessing in that it helped maintain some semblance of what our lives had been in times past, and it gave us short breaks from each other's growing despair.

Then, out of the blue, a call came from Jim Lyons. Without fanfare, he informed me that I was President Clinton's choice to be the next chief of the USFS. I thought it likely that Jim had had much to do with the president's decision. I was flattered and most certainly honored. But I told Jim that I simply could not accept the appointment for four reasons. First, Margaret's health was deteriorating ever more rapidly now, and her death was inevitable—maybe just weeks in the future. Second, I had no ambition to be chief. Third, I did not consider myself qualified for the position. And, fourth, I was not a member of the Senior Executive Service, a cadre of public servants who are groomed for career appointments as managers and executives of federal programs and agencies. If I accepted the position, I would be a political appointee and could expect a certain amount of flak. (I had previously avoided the training that could have led to SES membership because I would then have been obliged to accept any position in the SES to which I was assigned; I was not willing to make such a concession.)

Jim said that he hoped I would reconsider—but he understood and sympathized with my decision. I did not tell Margaret about the call.

Several days later, I came in from a day in the field to face a very irritated wife. Why, she asked, had I not told her about being offered the position of chief forester? Her dark brown eyes were piercing. "You've never before made a decision involving our lives without consulting me! Why now?"

Clearly, she had talked to someone in Washington—I never knew whom. My lady fair was fired up, and nothing but straight talk would suffice. For the first time, we discussed the inevitability of her death. She had told me, in writing, exactly what I was to do when that time came. We had agreed that, if we could control matters, she would die in her own home. She wanted to be surrounded and cared for by family and friends. She said that, we were, after all, partners "until death do us part." We didn't take those words lightly during our wedding ceremony, and nothing had changed that over the thirty-six years since.

She looked straight at me, brown eyes flashing. "First off, I don't want to dwell on where I am going to die. In fact, I don't even want to talk about dying. I want to talk about living—every single day that I have left. I have entertained our friends and your colleagues around my dinner tables for thirty-six years. I've listened to y'all talk, and talk and talk and talk, about how things could be better for those things most important to you. I gave up some of my dreams for your dreams and then our dreams. Now you have a real chance—the best chance ever—to help make those dreams come true. So now you back away? Worse than that, you gave up on your dreams, our dreams, without even talking to me about it."

I tried to protest. She cut me off by raising her hand in a manner that clearly said "stop." After taking a moment to recover some strength, she went on. "You may be afraid. But I'm not. If you turn away, you are doing it for reasons that don't involve me. I am ready to go to Washington. I want to go! We both know where your duty lies. You're scared and emotionally exhausted. I understand that, but you can't let fear or weariness stop you. Face the fear. Then do as you think best. But whatever you do, *don't you dare hide behind my skirts!*"

I have never been as proud of anyone as I was of Margaret at that moment, or more ashamed of myself. I never loved her more. I should have come to her with the offer, and then we could have made the decision together. Our eyes locked for a very long time, it seemed. Then the tears came, but not from her eyes. I did not speak but only nodded. We had come to a really big fork in the trail, and we chose our path together—whatever remained of it.

The next day, I called Jim Lyons and said I would accept the position. But that did not make it a done deal. The new administration had hundreds of jobs to fill. Each candidate had to go through a vetting process involving security clearances and background checks—and only heaven knew what else. Time marched on.

Finally, word came that we were to come to Washington for my swearing-in ceremony. Margaret was weak but eager to go. But once in Washington, we were informed that the FBI had fallen behind schedule on vetting my appointment. We lived nearly two weeks in a hotel—waiting, waiting, waiting as Margaret grew weaker. Then it was time for us to go home to La Grande.

As we changed planes in Salt Lake City, I was pushing Margaret in a wheelchair when I heard myself being paged. Undersecretary Lyons was on the phone to let us know that, by order of the president, the FBI had given my case priority and I had been cleared. We were to return to Washington for a swearing-in ceremony tomorrow.

That was out of the question. Margaret was pale and trembling with fatigue. She simply needed to get home. Jim insisted, but his heart wasn't in it. On the other hand, I was adamant. We were going home. Reportedly, my decision was not well received by the new secretary of agriculture. So Margaret and I never had a swearing-in ceremony. I regretted that. Margaret, I think, would have enjoyed the pomp and circumstance.

Two weeks later, we returned to Washington. Undersecretary Lyons, who knew Margaret from earlier and better times and understood her sacrifices, asked me if he could do anything to make things easier or more pleasurable for her. I told him that I thought she would cherish a meeting with the new president and vice president. He nodded but made no promises.

When we checked into our hotel, the clerk handed me an envelope, with the presidential seal in the upper left corner, addressed to Margaret. The enclosed handwritten note read, "Dear Margaret, I hope that you and Jack can visit with Vice-President Gore and me in the Oval Office tomorrow at 3:00 P.M. Undersecretary James Lyons can provide you with details. Respectfully, Bill Clinton."

Lyons, by God, had made it happen! His, and President Clinton's, gift to Margaret was in her hand.

At the appointed time, we were escorted to the West Wing of the White House, where we were met by Lyons and Katie McGinty, director of the Council on Environmental Quality. Margaret insisted on leaving her wheelchair behind, and we were ushered into the Oval Office to be greeted by President Clinton and Vice President Gore. For the next half hour or so, the president and vice president talked exclusively with Margaret as I stood by her side. I was so focused on Margaret's still-lovely face and bright eyes that I remembered very little about the conversation. I was struck, as always, by her erect bearing, charm, and the beauty of her very soul. Even worn down by over a year of illness and the ravages of chemotherapy, she was radiant. Now, though, her considerable beauty was more in her character, her eyes, and her bearing. She sat primly on the edge of the couch in the Oval Office, ankles crossed, her hands folded in her lap, alert and animated. She was not shy or intimidated in the least.

At the end of the meeting, President Clinton asked if she would like to have pictures made with him and then with the two of them. She smiled and nodded her appreciation. She stands erect in those pictures—thin, tired, and worn, but proud, beautiful, and erect. I will always be grateful for the honor paid her that day so long ago by the President and Vice-President.

Margaret was nearly spent on the ride back to the hotel. I asked the bellman for a wheelchair, and she didn't argue. When we got to our room, I helped her lie back on the bed and removed her shoes. She removed her wig—nobody but me ever saw her bald. I kicked off my shoes, lay down beside her, and held her hand.

Though exhausted, she was too excited to even doze. We lay quiet for some time. Finally, she said, "It has been a long and wonderful journey—Sonora and Llano, Texas; Morgantown, West Virginia; Amherst, Massachusetts; La Grande, Oregon; and now Washington, D.C."

We were partners. The homes she made in those wonderfully varied places had quickly become a gathering place for family, friends, and my professional colleagues. She raised two fine sons and took in other children who needed her help and guidance. She hosted many dozens of my professional colleagues, some from faraway exotic places. She made no pretense of having expertise in natural resources—her bag was music and people. But she believed in me and in what my colleagues and I were trying to accomplish. After her death, I received several notes from colleagues who referred to her as "a first lady of American conservation." She would have loved that sentiment—and the title.

*Margaret and Jack at a specially arranged meeting with
President Bill Clinton, Washington, D.C., 1993.*

She spoke again. "You are where you should be, with a chance for you and your friends to make a difference." She took my hand in both of hers, kissed it, and went to sleep.

Over the next six weeks, we settled into our new home, a condominium in Arlington, Virginia. She slept a little more each day. We talked but little; nothing had been left unsaid. Instead, we cuddled, held hands, listened to her classical music favorites, and waited. Oddly, I remember those days as good days—even as the best of days. The photographs of her with the president and vice president, embellished with their handwritten good wishes, sat on her grand piano, the ultimate place of honor in our homes.

In her last days, she stirred herself from increasingly deep and longer periods of sleep to say good-bye to her children and her first grandchild. We were alone in the small hours of the morning when she awoke for the last time and weakly squeezed my hand. She said, "It's a long way . . . ." Her voice trailed off.

After a time, her ever more shallow breathing ceased. I held her hand for a while and then awakened our younger son, Greg, who had been so faithful and helpful. I gave him the numbers to call to have Margaret's body taken to the crematorium. I was too exhausted—and I simply couldn't bring myself to perform that last service. It was a blessing—for Margaret and for me—that Greg was there.

I flew from Washington to Pendleton, Oregon, with her ashes in a box that I carried on my lap. My old friend and colleague Larry Bryant met me at the airport and drove me over the Blue Mountains to La Grande. The Methodist church in La Grande, where she had been the choir director for some twenty years, was filled to overflowing for her funeral service. As I stood in the cemetery watching her ashes being placed in a crypt (which had a place reserved for my ashes when the time came), I looked across the Grande Ronde Valley to the Eagle Caps and thought of our many trips there together. It made things a little easier.

Surely, it had been a long journey for Meg from the tiny oil-patch town of Monahans in the Permian Basin of Texas to Washington, D.C. But her journey was not yet over, for it would continue so long as I lived. She has been, and would always be, part of me and I part of her, and we would travel together always. Jumbled memories and thoughts cascaded through my mind as I sat there holding her hand. Among them were the words she had spoken to Kath Ann Feltis so many years before.

"Honey, you won't be lonesome. You have hundreds of friends there. Some you know and hundreds you haven't met yet—hundreds."

I hoped it was true. I knew that Margaret knew it was.

ADDITIONS TO THE HALL OF HEROES

Unlike my twelve predecessors as chief of the USFS, I had never had the ambition to be a full-time administrator. Several times I had declined opportunities to undergo the training to become eligible to join the elite cadre of the federal government's Senior Executive Service. My predecessors were identified early and carefully groomed to be chief by a combination of appropriate training and strategic placement in a series of line and staff assignments, including a tour or two in the Washington, D.C., headquarters.

I had chosen a career as a research scientist. As a result, I spent the first twenty-seven years of my USFS career in the "boondocks" (that is, outside Washington's Beltway) doing research and leading research teams. Over the years, I had been privileged to lead a series of high-powered research teams dealing with extremely controversial and highly significant issues in conservation. The skills of my colleagues and the results we produced reflected well on me.

My years as a USFS researcher ultimately earned me the highest possible rank (Science-Technical Grade 17—ST-17) that could be attained. It was an exceedingly rare honor. To make my surprise appointment even more puzzling, I was the first chief who was not a forester or engineer—wildlife biology/ecology was my bag. My successor was to be a fisheries biologist. Likely, those appointments reflected changing times and changing priorities.

I arrived in Washington to take on my new job on a Friday in late December of 1993. On Saturday, Margaret asked if I would take her on a tour of the USFS's headquarters building and my new office. Her cancer was in remission and she was, for the moment, bright-eyed, bushy-tailed, and flushed with excitement.

When we entered the headquarters building, we were stopped by security guards who had no reason to know me by sight. It took a while to explain who I was, prove it, and gain entry. While the guards were embarrassed, it was an amusing and appropriately humbling experience for me. Our first stop was my new office on the third floor. Despite its glorious view of the Washington Monument, the office was large, imposing, and sterile. Every sign of the previous occupant, Chief F. Dale Robertson, had been removed. I knew the same would be true upon my departure, the only question being when and under what circumstances. But that lay in the future.

The desk that the first chief, Gifford Pinchot, and most of his successors had sat behind was obvious by its late nineteenth-century style. A USFS legend had it that Chief Pinchot left a letter of instructions in that desk detailing actions to be taken should the existence of the USFS be threatened, especially by transfer to the Department of the Interior. Just in case, I carefully looked through the desk and found no such letter, not even a secret compartment.

I felt self-conscious and a bit intimidated when I sat down behind "the Pinchot desk." Margaret smiled and said, "It will fit just fine with a little wear—like a new saddle. I'll bet every chief who sat down behind that desk for the first time felt the way you feel right now. If they didn't, they should have."

The next stop was the Chief's Conference Room on the second floor. The room was reached via a long, imposing hallway with high ceilings. The walls were bare save for the framed pictures of all the former chiefs, adorned with plaques inscribed with their names and the years of tenure. Individual spotlights were focused on each picture. As I studied the portraits, I sensed that something was missing from that "hallway of honor."

Margaret wearied quickly, and too soon it was time for us to go back to the hotel. That night, as we lay in bed talking, I told her of the nagging feeling that had come over me in that hallway.

She said, "I know that there were USFS heroes who were never chief but who were responsible for big lasting changes in how the national forests were managed. Maybe it's the pictures of those heroes that are missing from that wall of honor?" Margaret had intuited the answer. She often said, rightly, that I tended to think too much and feel too little.

My first item of business on my first day as chief was to meet with the deputy chiefs and associate deputy chiefs in the Chief's Conference Room. I gave my first instruction. I wanted additional portraits hung on the wall opposite the pictures of the chiefs—pictures of agency heroes who had led, or even forced, the agency to think "outside the box" and expand the value and potential of the national forests as fish and wildlife habitat, grazing lands, watersheds, wilderness, and recreational areas.

I had my list in my breast pocket. First was President Theodore Roosevelt, who along with his friend Gifford Pinchot dramatically expanded the USFS in the face of determined opposition from powerful members of Congress from the western states, most from his own Republican party. He backed Pinchot in creating a professional workforce of natural resource managers under the Civil Service, thereby denying subsequent chiefs the ability to dispense political patronage.

Aldo Leopold, a leading proponent for the creation of wilderness areas, beginning with the Gila Wilderness in 1921, was second on my list. He was trained as a forester at Yale, partially on the Pinchot estate (Grey Towers) in Pennsylvania. He spent his formative years in the USFS in the Southwest. Though trained and credentialed as a forester, he went on to become a professor of the new field of wildlife biology at the University of Wisconsin. His philosophical writings—*A Sand County Almanac*, paramount among them—were a seminal force in raising the nation's environmental concerns. He assumed a high-profile role in the establishment of The Wildlife Society, the Ecological Society, and The Wilderness Society. He came to be recognized as the father of wildlife conservation in North America.

Robert "Bob" Marshall was the key player in stimulating the USFS to recognize the burgeoning demand for outdoor recreation and the role that the national forests could play in satisfying that demand. He was a force in the wilderness movement and a founder of the Wilderness Society in 1935.

Arthur Carhart was the USFS's first landscape architect. Though he served in the agency for only three years (1919–1922), he was instrumental in designing and implementing plans to meet growing demands for backcounty recreation, starting with the establishment of the Boundary Waters Canoe Area in Minnesota.

When I finished my presentation, there was no wild applause from the senior staff, but there were no objections either. I thought to myself, being chief is good! My very presence in the chief's job and the new pictures that would appear in the "hall of heroes"—mine included—were symbolic of change. I could only hope it was change for the better.

The "writing on the wall" was meant to convey a message— the agency's future would entail less emphasis on timber production, livestock grazing, and minerals extraction and increased attention to fish and wildlife, recreation, and watersheds. That was to be achieved through the emerging concept of ecosystem management. My immediate predecessor, Chief Robertson, had declared such as a hugely symbolic shift in USFS policy.

The plethora of environmental laws passed in the previous three decades and interpreted by the courts had fostered that recognition and shift in policy. In my opinion, those pressures would only intensify, and there would be no turning back. It was my intent to embrace the future by recognizing USFS stalwarts from the past and to formally declare them "heroes" to be emulated.

At the end of my tenure, when I hosted the press for the last time in the Chief's Conference Room, I spoke briefly, answered a few questions, and left. As I walked along the hallway, I looked at the pictures of Roosevelt, Leopold, Marshall, and Carhart and wondered if they would still be hanging in that place of honor in a decade or so. That would, I thought, be a legitimate test of my tenure.

In early 2005 the USFS celebrated its 100th birthday with a big bash in Washington. All the living former chiefs—R. Max Peterson, F. Dale Robertson, Michael Dombeck, and I—were honored guests of the current chief, Dale Bosworth. I was visiting the USFS headquarters for the first time since my departure nine years earlier. This time, the security guards recognized me and ushered me in. I walked the hallway to the Chief's Conference Room on the third floor to see if the pictures I had ordered hung were still in place.

The photos were hanging exactly as I had seen them last. On the opposite wall were three additional pictures of the subsequent chiefs since my time there. Maybe there was a message in that. It came to me that the greatest honor that any chief could be paid would be to have her or his portrait hang on both walls. Now, that would be something!

WHERE'S MY PICTURE?

In January 1905, the Forest Reserves were transferred from the Department of the Interior to the Department of Agriculture and renamed National Forests. The USFS was simultaneously created to protect and manage those lands, fulfilling a vision of President Theodore Roosevelt and his partner in conservation, the first USFS Chief Gifford Pinchot.

Pinchot believed that the Forest Reserves, authorized in 1891 and set aside from the lands in the public domain, should be retained in public ownership, which would be possible only if the lands served the people by being carefully and prudently managed to produce goods and services, including jobs. That would be achieved primarily through carefully regulated grazing, timber harvest, and extraction of minerals. He believed it was critical that management decisions be made by technically competent, locally assigned USFS officers on local grounds.

Pinchot coined the phrase "wise use" and stated that the national forests should be managed to provide "the greatest good for the greatest number in the long run." That approach, he believed, would be more likely to be sustained in the Department of Agriculture than in the Department of the Interior, which during that period was often accused of corruption. I had my own—very different—suspicions. Given his patrician ego, I suspected that Pinchot wanted as little supervision/oversight as possible. Secretaries of agriculture could be counted upon to devote most of their attention to agricultural programs and associated politics. That would leave little time for issues related to management of the national forests.

If so, the chief of the USFS would have more of a free hand—and hence more power—relative to other agency heads. The century following the creation of the USFS in 1905 was marked by continuous tension over the degree of independence of USFS leaders, tensions that waxed and waned depending on the strength of personality of the individuals involved and the circumstances of the moment.

Secretaries of agriculture became more and more aware of the USFS as the timber cut from the national forests soared between 1945 and 1990. A modified version of the "golden rule" came into play, as might have been expected. The more "gold" that was associated with an agency's activities—and the more controversy—the more attention it would be paid by elected officials. Money, politics, and controversy commonly moved in lockstep.

On average, over half of the USDA's employees are in the employ of the USFS. The president usually appoints the secretary of agriculture from a "big agriculture" state, and the secretary focuses attention on such issues as crop subsidies, international trade, food stamps, agricultural research, extension, and producing political support for his boss, the president. Most secretaries of agriculture have known little—and cared even less—about the USFS and the management of the national forests. Conflicts related to the USFS and forest management were commonly viewed as a diversion from what most secretaries of agriculture have considered more important issues.

Historically, then, USFS chiefs have been neither inclined nor invited to spend much time hanging around the agriculture secretary's office. During my tenure, the secretary's routine staff meetings for undersecretaries and agency heads almost never dealt with anything remotely related to the USFS. During the latter part of my tenure, the secretary of agriculture paid little attention to me or to the USFS, which suited me just fine.

Exceptions arose, from time to time, when Secretary of the Interior Bruce Babbitt—usually in cahoots with Katie McGinty, chair of the Council on Environmental Quality—"poached" on USFS matters to mollify the Clinton administration's carefully cultivated environmental constituencies. Then, at unexpected moments, the agriculture secretary would reengage, assert his authority, and then once again quickly fade into the background.

During one of those "reengagements" the agriculture secretary paid a visit to the USFS's offices. When he arrived in my office, he was obviously irritated and wanted to know, "Why is my picture not hanging in the entry hall?"

Somewhat puzzled, I replied that his picture was hanging in the lobby.

"No," he replied. "Your picture and President Clinton's picture are prominently displayed. But, typical of the USFS, my picture is nowhere in sight. You and your people report to me and not to the president."

I assured the secretary that I would check into this obviously very important matter. The rest of our business was conducted with tension in the air. When the secretary departed, he insisted that I accompany him to see for myself that his picture was not on display. He was right. The place where his picture ordinarily hung was vacant, with only a hole in the masonry wall.

I joked, "Maybe a member of your fan club stole the picture?" The guards at the entry desk laughed. I laughed. Neither the secretary nor his two staff assistants smiled.

Clearly, the USFS and I had a new first-priority problem! All hands on deck!

Keen detective work, which took less than a half hour, revealed that the framed photograph of the secretary, in the process of being dusted, had fallen to the floor two days earlier when the picture hook had pulled out of the very old masonry. The glass and the frame had shattered and were being repaired.

How's that for luck? The secretary paid one visit to the USFS's office in two years, and his picture was not hanging in the lobby. And then he actually noticed—and actually gave a damn. I was impressed—how was another matter.

I called the secretary's chief of staff and explained the situation. To prevent the recurrence of such a serious *faux pas*, I said I had ordered that the secretary's picture be prominently displayed in two places in the lobby. And, of course, the secretary's picture would be twice the size of the chief's picture—and just a smidge smaller than that of the president.

I sensed that the chief of staff was struggling to keep a straight face. He said he would convey my abject apologies and my solution to the crisis to the secretary. And, of course, he would convey my heartfelt apologies. Okay, with that taken care of, we could get back to work.

The secretary of agriculture set foot in the USFS office only once more during my tenure. He presided at the press conference held on the day of my departure. I had learned a valuable political lesson. No matter how trivial it might seem, take care of the little things—especially those that might involve the stature and ego of superiors.

NOT POLITICALLY CORRECT?

Beginning in the 1980s, the federal government was swept up in a growing wave of concern over civil rights and launched efforts to assure that qualified persons who were underrepresented on the basis of race and gender were recruited and then treated equally. The worthy objective was a federal workforce that "looked more like America." Emphasis was placed on the accelerated recruitment, training, and advancement of underrepresented racial minorities and women.

Over its nearly 100-year history, the USFS had evolved to be staffed primarily by professions that were overwhelmingly dominated by white males: forestry, civil engineering, wildlife and fisheries biology, range management, firefighting, and law enforcement, among others. Further, most such jobs were located in the rural West, where some minorities commonly numbered between few and none.

That resulted in the natural resources agencies (including the USFS) being the most white and most male of federal agencies. Understandably, minorities and women were pushing hard for their place in the sun. Conversely, many white male employees, with some justification, were convinced that they were now at an increased disadvantage in terms of hiring, retention, and promotion. It was a state of affairs guaranteed to produce tension, conflict, and enhanced sensitivities. "Doing right" would be neither easy to define nor accomplish.

President Clinton had made civil rights issues a focus in his presidential campaign, and the pressures upon federal agencies in that regard were strong and consistent. Secretary of Agriculture

Mike Espy informed me that meeting civil rights goals was my number-one priority.

I had just emerged from three years of dealing with conflicts in the Pacific Northwest whose resolution had resulted in the loss of many thousands of jobs. To me, Secretary Espy's emphasis on "civil rights" did not seem to be quite on a par with that. But he held legitimate authority over the USFS and spoke for President Clinton. I essentially replied with an "Aye, aye, sir—I understand and will obey."

Four months after becoming chief, I convened a "family meeting" of USFS employees stationed in Washington, D.C. In the course of that meeting, I announced recent promotions and new assignments. In the process, I introduced Dr. Ron Stewart as my choice to fill a critical administrative position in the D.C. headquarters. Ron and I were friends after serving together as senior research scientists at the Pacific Northwest Forest and Range Experiment Station, and I mentioned that Dr. Stewart was uniquely qualified, having previously served as both a regional forester and a research station director. Then, in an attempt at humor, I looked at Ron's bald head, rubbed my own bald pate, and said, "In addition, I have always admired the way Dr. Stewart cuts his hair!" Ron led the laughter.

Two days later, the USFS's deputy chief for administration appeared in my office and came directly to the point. "Chief, do you recall your comments when you announced Ron Stewart's promotion?"

"You mean about his being both a regional forester and a research station director?"

"Well, I was thinking of the joke you told about the two of you being bald."

I nodded, "Yeah?"

"I have received two civil rights complaints based on that comment." He wasn't smiling.

"What? How could that statement generate a civil rights complaint? Surely Ron didn't complain?"

The deputy chief shrugged, sighed, and replied, "It is asserted that you indicated that, in addition to other factors, you selected Ron because he was bald."

"Yeah? Are we bald guys a protected class under civil rights protocols?"

"Chief, the complaints were filed by two different women who contend that you meant to convey, under cover of a joke, that you picked Ron because he was male. Since men go bald and women ordinarily don't, you conveyed a not-too-subtle coded message to the 'good old boys' that you will continue to favor males as you fill top jobs."

My jaw dropped. "Damn! My record in the area of civil rights clearly shows that I am a strong advocate for absolutely even-handed treatment in both hiring and promotion. That interpretation is paranoid, wrong, unjust—and stupid!"

The deputy chief said he thought he could get the complaints withdrawn, and I heard no more about the matter.

For several months after my wife Margaret passed away, some female coworkers commonly greeted me with a hug displaying their sympathy and affection. My civil rights advisors warned that such behavior could cause me trouble and that I should make it clear that hugs from employees—especially those of the female persuasion— were not welcome. I replied that I would not push away any coworker, female or male, who expressed affection or sympathy for me with a hug. I would risk the complaints. Attention to "political correctness," while legitimate, could go too far.

I called another meeting. I made it clear that I believed that all employees should be recruited and treated equally regardless of sex, ethnic background, sexual orientation, or age. But the developing atmosphere was destructive, distractive, and in some cases irrational. Further, I had no intention of turning my back on my culture. I would not attempt to change my Texican accent—I talked that way because of where I grew up. To hold that against me was unjust. I would not turn away heartfelt expressions of caring and concern. I made it clear that I would rather return to my research job in La Grande—or retire—than live my life as a sterile robot self-programmed to eschew emotion, caring, humor, and manner of speech.

I had expected, at least from some, a hostile reaction. Instead, I received a standing ovation. It was a turning point, as festering hostility among a few gave way to improved understanding that changes were justified, required, needed, and would occur with as much respect and understanding as possible.

I then announced that I had a well-deserved reputation (if I did say so myself) as a good storyteller and jokester. But, I promised, from then until the end of my USFS career, I would never again tell a joke—not even at my own expense. I had more important things to do than deal with complaints that bled time and money and attention from the agency's core mission. A humorless workplace, I said, was a high price to pay for a politically correct atmosphere.

The crowd booed; the ice was broken. So humor and storytelling remained my trademark.

It was a rough start with a good ending.

FULL STOPPER

In 1994 and 1995, Republican congressmen from the western states, where most of the national forests were located, made up the bulk of the membership of the House and Senate committees that dealt with USFS matters. The leaders of those committees were, by and large, seriously displeased that the timber yields from the national forests had declined steadily over the previous decade—and were still declining.

The brighter or more honest members realized, but studiously ignored the fact, that declines in timber yields resulted from adherence to interacting laws as enforced by the federal courts when coupled with reductions in appropriations and more "line items" in the budgets reduced agency flexibility. Many of those same members, now disgruntled, had voted for these very laws during the Richard Nixon administration and clearly suffered from "selective memory syndrome."

Most members recognized, but wouldn't publicly acknowledge, that the environmental communities were winning lawsuit after lawsuit. Each win forced enhanced efforts at compliance by federal land management agencies. That is the way the system is supposed to work. Congress, of course, could have changed the applicable laws if they didn't like the results.

Now it was becoming more and more politically advantageous for politicians to attack the USFS as it was performing under the

Clinton administration. The attempt was intended to mollify workers, corporations, small businesses, and other constituents hurt by ongoing declines in timber harvested from the national forests. In the 1994 mid-term elections, the Republicans attained majorities in both houses of Congress, and congressional committees dealing with federal land management agencies brought on hired hands ("staff") with strong ties to the timber industry.

In doing so, congressional leaders seemed to care more about showing their constituents that they were working to increase timber yields than they actually cared about the timber harvest. The political theater included holding a well-choreographed sequence of hearings in the Senate Committee on Natural Resources and the Environment and the House Committee on Natural Resources. The aim was to make the Clinton administration appear solely, or primarily, responsible for ongoing declines in the timber yields from national forests. I came to believe that the hearings were intended to set the stage for a revision of the tangled mess of laws and mismatched court decisions afflicting federal land management, which I thought was long overdue.

The stagecraft for the committee hearings emphasized the power and grandeur of the Congress—and, indirectly, the power of the members—relative to that of the witnesses. The members sit on a raised horseshoe-shaped dais with several tiers of seats looking down on witnesses. Witnesses sit at a table down "in the pit" looking up at the members. The ceilings of the hearing rooms loom large. Lights are focused so that the witness feels the pressure.

The circumstances could be intimidating to the uninitiated and sometimes amusing to veterans of political gamesmanship. I came to believe that it was the public, especially constituents, that the stagecraft was intended to impress.

While trying to be truthful and supportive of the Clinton administration's positions (with which I sometimes disagreed), one of the "games" I devised was to be alert for opportunities to slip in a "stopper"—a polite, technically correct answer to a question that left inquisitors with little or nothing to say. It added spice to the exchanges. Since the inquisitors have all the advantage, a stopper is difficult to deliver.

To be effective, a stopper must be carefully timed and delivered—like springing a wolf trap. No hint of sarcasm or disrespect should be detectable in the words, tone, or expression of the witness. Government witnesses, especially, must appear respectful and courteous, regardless of their sometimes ill treatment—at the risk of being held in contempt. Alas, that requirement does not extend to members, some of whom appear as arrogant bullies and make fools of themselves with some degree of regularity. The "playing field" between the members seated on the dais and the witnesses is not a level one—not even close. The deck is stacked for or against the witness depending on the wishes and actions of the committee's chairman.

During my first year as chief, Democrats held the majority in both the House and the Senate. The hearings were contentious, but the chairs of the committees assured that I was treated courteously. When Republicans won the majority in both houses, the atmosphere flip-flopped.

The next several hearings in which I was a witness were, to my mind, outrageous. They included thinly veiled abuse and overt badgering of administration witnesses. The effect was that many, sometimes all, of the Democrats on the committees I dealt with simply quit showing up for hearings. Even some more moderate Republicans shied away, depending on the subject of the hearings. Witnesses who represented federal land management and conservation agencies were all too often under attack from opening to closing gavels. I was reminded of the "bread and circuses" during the glory days of the Roman Empire to entertain the masses. Whatever the purpose of the hearings, they were, in my opinion, not the way to solicit information or search for answers. Most of the hearings I was party to were pure theater.

I especially remember one hearing in front of the House Committee on Natural Resources. The hearing, which nary a Democrat attended, was long, drawn out, and acrimonious. Every Republican member took the opportunity to perform for the nightly news programs in their home districts. Toward the end of the hearing, a very young just-elected congressman leaned forward to his microphone, peered down at me disdainfully over his reading glasses, and

Jack's official portrait as USFS chief.

asked in a sarcastic tone, "Chief Thomas, just who do you think you work for?"

I explained the chain of command: president, secretary and undersecretary of agriculture, down to me.

The inquisitor looked down and sneered, "Really? Don't you think you ought to work for the people of this country—the American people?"

It was a good cheap shot. I replied, "Yes, in the final analysis, I work for the people of the United States—as described in law and budget direction."

He bored in, "Well, Chief, do you work for the thousands of workers in the timber industry who lost their jobs because of you, those families now on welfare because of you?"

I answered something like this: "Yes, sir. I work for those people. They are part owners of the national forests. I work for all the people, including those yet to be born. The chiefs who preceded me did their best to do that for me—and for my children. Now I should do no less. But we must obey the applicable laws, passed by Congresses, signed by presidents, interpreted by various federal court decisions, and within budgets passed by the Congress. If and when Congress changes those laws, we will comply."

The chairman, with a scathing glare, shut me and the hearing down. My antagonist, which was the way I now thought of him, made a dramatic exit. I did not look forward to the next encounter. I knew he would be better prepared next time.

Damn, there were days when I loved that job! Triumphs for witnesses are rare and fleeting. Consequences come later.

NATURAL RESOURCES POLITICS, ALASKA STYLE

The declines in timber yields that began under President George H. W. Bush were continuing under President Clinton (and declines would continue apace under President George W. Bush). It didn't seem to matter what president was in office or which political party was in power—the declines in timber yields inexorably continued. Though many elected officials railed, they were unable or unwilling to change the laws that produced the situation.

Collectively and over time, the politicians had largely produced the ongoing mess but looked everywhere else to place blame. Many politicians at both federal and state levels now found it politically expedient to attempt to scapegoat individuals in the government (the "bureaucrats") who were carrying out the laws as passed by Congress, signed by presidents, and interpreted by the federal courts.

Pointing fingers was much easier than making adjustments in the applicable laws—and their interactions—that had mandated the ongoing changes in federal land management. Politicians, in the time-honored way of uniquely American politics, worked hard to "put the turd in someone else's pocket." I was gaining an understanding of the ongoing game and developing some applicable skills, but I didn't like the game. It should have been insulting to the intelligence of the American people, but the vast majority did not even know that such a game was under way. If they did, they didn't seem to understand it.

We scientists who had been assigned to deal with the spotted owl/old-growth issue had become the inadvertent poster child for those opposing changes in forest management that were necessary to comply with the interactions of the environmental laws (especially the Endangered Species Act and federal court decisions). That identity was to follow me when I became chief.

The three members of Alaska's congressional delegation stood out in the midst of the accelerating political wrangling. Alaska contained 22,750,000 acres of national forest, or 12 percent of the total. The Tongass and Chugach National Forests contained most of the federal lands in Alaska capable of producing commercial timber in a sustainable fashion. Harvesting and processing timber had sustained, by far, a larger proportion of Alaska's workforce than that of any other state.

In 1993, when Republicans were in the minority in Congress, Alaska's congressional delegation was long on seniority but short on power. Senator Ted Stevens was the senior minority member on the Senate Appropriations Committee. Senator Frank Murkowski was the senior minority member of the Senate Committee on Energy and Natural Resources, and Congressman Don Young was senior minority member on the House subcommittee that dealt with USFS matters. I referred to them as the *"tres amigos."* Their ability to vent their frustrations with the USFS was only somewhat restrained by their minority status. They were eagerly looking forward to changing that in the next election.

The three were, understandably enough, distressed with the steadily declining timber yields from the Tongass and Chugach National Forests, which fed raw materials to two Japanese-owned mills. Those mills were the product of two long-standing sweetheart deals: two fifty-year timber sale contracts that stood out in contrast to the competitive bids required of other producers of wood products. Such a deal may have made sense just after the close of World War II, when Japan needed raw materials to rebuild and the United States wanted to build the population and economy of Alaska. But now that time—and its preferential treatment for Japan—had long passed.

Alaska politics are somewhat less polite than in most states in the Lower 48. The Alaska delegation routinely began their political discourses—public and private, with me and other USFS staff—with a figurative knee to the groin and an elbow to the nose, followed by a thumb in the eye and a chomp on an ear. And that was just to get one's attention. I got my first taste of what lay ahead in mid-January of 1995, shortly after Congress reconvened and Republicans were in the majority.

While making my get-acquainted rounds of the offices of the senators with whom I would be dealing, I was buttonholed by one of Senator Frank Murkowski's aides, Greg Renkes. He informed me that if I did not hew to the Alaska senators' desires relative to USFS activities in Alaska, I was "in line for some serious ass-kicking."

I smiled and said, more confidently that I felt, "Pack a lunch. That might take a while."

He smiled—in a way that seemed more of a sneer to me—and walked away. I knew that the two years until the next election would be, at the very least, interesting. The Alaska congressional delegation held the chairmanships of all three committees with the most influence over USFS affairs.

After a while, the Democrats on those committees simply quit showing up for meetings. Whether they were shell-shocked, didn't care, didn't know how to behave in a minority role, didn't see any point in trying to influence committee decisions, or were simply disgusted with the situation was never clear to me. But the Democrats on those committees, with a few exceptions, could have been appropriately classified as "missing in action" during 1995 and 1996.

My first hearing after the new Congress convened took place in front of Congressman Don Young's committee in the House. The chairman strode in, took his seat, and looked around the room. He stood out from his colleagues. He looked exactly like what he had been in "real life"—a tugboat captain with weathered face, bit of scraggly gray beard, a bad comb-over hairstyle, and an unfailingly straightforward demeanor.

Twelve Republicans were arrayed to his right. No Democrats were seated to his left. He rapped his gavel, smiled his best Cheshire-cat smile, and peered down at me. "Well, Chief, I see twelve angry Republicans and no Democrats here today. I suspect you are in for a long, hard day."

I leaned forward to the microphone and asked permission to correct the record. Mr. Young look puzzled but nodded permission. I could not resist being a smartass under the cover of humor. "Mr. Chairman, counting you, there are thirteen Republicans seated on the dais. I will take your word for it that they are 'mad.' If our subject is confined to the U.S. Forest Service and natural resources management under applicable laws, I figure that, on balance, I can hold my own. Actually, I am looking forward to some intelligent give and take."

Recognizing the oxymoron, some in the audience laughed, though in a subdued fashion. Most of the Republicans seated at the dais seemed bemused by my audacity. Several laughed; some seemed puzzled. I kept smiling, thinking of the Cheshire cat. Two can play at that game. For several hours, the committee members, one after the other, came after me as if I were the devil incarnate.

I was relieved to discover that I was not at all intimidated. Most of the "questions" were really statements. My staff had prepared me well by anticipating questions, attacks, and statements and by providing me with detailed responses or data that I could use at my discretion. And, after all, what would be discussed was "my bag"—at least from the standpoint of education and four decades of experience in the natural resources arena.

Given their belligerent attitudes and their treating me as "the enemy"—as opposed to an expert witness—I moderated my replies to hostile questions while trying to appear at least marginally respectful. I tried to answer their questions, and to their statements in the form of questions, as if they made good sense. Many of their questions would have failed that test. Pretty soon the audience's titters, sometimes punctuated by outright laughter—sometimes from members of the press—indicated that this first hearing was not going quite as planned by the committee's inexperienced staff.

When Chairman Young gaveled the meeting to a close, I approached the dais and asked if we could visit for a few minutes in private. He nodded, and we headed for his office as soon as we could free ourselves from the reporters waiting in the hall. We knew each other from membership in the Boone and Crockett Club.

Though he was irritated, out of courtesy, he gave me the first word. "Mr. Chairman, your committee set out to embarrass or intimidate me—maybe both. In my opinion, all concerned should be embarrassed, the two of us included. I am a professional conservationist with nearly four decades of experience. I am not a political partisan, having been both a registered Republican and Democrat over the course of my voting life.

"I accord you respect as an elected official and as chairman of this prestigious committee. I will treat you with respect in public, no matter how you treat me. But I won't sit quietly and be talked down to or subjected to ridicule.

"When you deal with natural resources, you are playing my game on my field with my ball. I am well-trained, overeducated, and thoroughly experienced. I eat and sleep natural resources and have done so for going on four decades. I have a staff that briefs me thoroughly before I come in front of your committee. They sit just behind me during hearings to give me any information I need when I need it. They are skilled and experienced. I don't believe you can best me in anything resembling a fair game of give-and-take. And if it becomes too blatantly unfair, I will make that obvious to the press in the room.

"Natural resources and their management are critically important matters; we both understand that. I will provide you any information that you request—accurately and with dispatch. If you treat me
and my fellow USFS professionals, many of whom are your constituents, with the respect we deserve and have earned, we will do our best
to answer your questions fully and quickly. On the other hand, if you
come at me again like you did today, I will reciprocate to the best of
my ability—in a calm and measured fashion. You don't scare me. In
fact, I find blustering and threats amusing. I will do better next time
if you will do the same."

He smiled and nodded; he was, after all, an old political street
fighter. We shook hands. From that time forward, exchanges were
markedly improved, if not always gentlemanly. Over time, Chairman
Young and I developed a mutual though grudging respect. I thought
that neither of us were what we seemed in our three-piece suits and
shiny wing-tip shoes. Underneath all the posturing he was an old
tugboat captain. And underneath all the bullshit and fancy title I was
a field-oriented wildlife biologist more at home in a small town in the
West than in Washington.

That became obvious a year later, when several senior administration officials, including a member of the President Clinton's
cabinet, several undersecretaries, and agency heads (including me),
testified jointly before Chairman Young's committee. In an unusual
move, Chairman Young instructed that the witnesses stand and be
sworn. It was a calculated and obvious insult, intimating that truth
would not otherwise emerge.

Then Chairman Young looked down at me and said, "Chief,
there is no need for you to be sworn." *What?* Those words either recognized my professionalism, or respect for the position of the chief, or
maybe for the USFS. It was, in left-handed fashion, high praise indeed.

Then it was the Senate's turn. Senator Frank Murkowski had
hired a former top-drawer lobbyist for the timber industry as his committee's new chief of staff. He was one of the brighter and hard-working folks in the natural resources arena and thoroughly understood
both the technical and political facets of the natural resources business, especially as it related to management of public lands. This hear-

ing would not be a repeat of the "amateur hour" that marked the hearing in the House.

With opening statements out of the way, the hearing became a wide-ranging grilling of the USFS chief. Senator Murkowski, when not posturing as a hardass Alaskan, seemed to be a gentleman with a countenance to match. Quite obviously, he was not well versed on issues related to management of the national forests.

So the aide, seated behind the senators, often leaned forward and whispered what he considered the appropriate questions to be asked. Murkowski leaned back to listen to the aide and then leaned forward to the microphone and repeated the aide's question as if it were his own. As the process went on, it became increasingly obvious that it was the aide who was really asking the questions and providing leadership for the committee. I couldn't figure out why the aide didn't simply place a list of questions in front of Murkowski so that the senator would look somewhat less like a puppet on a ventriloquist's knee. I suspected the aide wanted to play the game with maximum flexibility by changing his questions, depending on my answers. He knew, I surmised, that I had a good chance of finessing his questions if I could discern a pattern. It was becoming more and more difficult not to be fascinated by the game itself.

Once, when Senator Murkowski leaned back to get his next question, I asked if I could make a statement. The senator nodded. I said, "Senator, I know that, having just taken over as committee chairman, along with your other duties, you are a busy man with many demands on your limited time."

The Senator nodded solemnly—he obviously concurred that he was indeed a busy and powerful man with many demands on his limited time. I continued in a solicitous tone, "Perhaps, Senator, we can spend the rest of our time today answering your own and the other senators' questions? I will be happy to meet with your aide and answer his questions at a mutually convenient time."

Laughter rippled through the hearing room. The aide retreated to scribbling notes and slipping them to the senator, who had trouble deciphering the handwriting. As a result, he asked several less than fully cogent questions. The fumbling afforded me opportunity to restate the question, for purposes of clarity, of course, and make points that I wanted to make.

Senator Murkowski called on Senator Larry Craig from Idaho. Larry Craig was too smart, shrewd, and experienced to be so easily thrown off his game. We got along fine outside the hearing room and had for several years. But nonetheless the hearing continued to lose focus. My impression was that Senator Murkowski was not pleased. I expected that the aide would see to it that the committee's "dance" was better choreographed next time. I was not to be disappointed.

Senator Ted Stevens summoned me and several of my top staff—including the USFS's regional forester in Alaska, Phil Janik—to meet with him in a basement conference room in the Capitol Building. With just the appropriate amount of tardiness suitable to a senator of his seniority and power, Senator Stevens strode into the room with several staffers in tow. We stood out of respect.

He sat down directly across the table from me, ignoring my hand outstretched in greeting. Senator Stevens, a short and slightly stooped man, seemed to me to be perpetually petulant and angry—maybe because he was. He glared at me. "Does the name George Dunlop mean anything to you?"

I was puzzled. "Yes, sir, I believe that he was once undersecretary of agriculture with responsibility for the USFS."

"Bingo! Do you remember anything about his relationship to me?"

Where was this going? "Yes, sir, if I recall correctly, you instigated action to cut off his salary through the budget for Interior and Related Agencies."

The senator's gaze never wavered, "That's exactly right. If I were you, I would keep that in mind."

I leaned forward and asked, "Senator, did you just threaten me? If you cut off my salary because I won't accede to your demands, you would simply make me a hero to many folks—at least to the folks whose opinions matter to me. I would be happy to return to my work as a research scientist." I was smiling, but just a little, not wanting to appear overly disrespectful or aggressive.

But I hoped my message was crystal clear: "Stuff your threat where the light don't shine—sideways!" He blinked first. I continued. "Senator, I'm no fool and you are a powerful man, one who can exert influence over the USFS and its operations in Alaska. If I can accommodate your requests, both legally and ethically, I will, of course, see what I can do. But threatening me, or the USFS, whether directly or indirectly, is not apt to be productive for you, the USFS or Alaska."

That threw the old bully off track, and it showed. The meeting ended with his general complaints about ongoing planning activities for the Tongass and Chugach National Forests in Alaska. I knew, upon parting, that I had not made a friend of one the most powerful of Republican senators.

I thought back to when I was ten or eleven years old and a bully ran me home from school. Dad was working out in the yard and saw what was going down. He made it clear that, even if I got my ass kicked, I was never to run away from a bully again. The next day I punched the bully, who was a bit bigger than I was, and that was the end of that. Was this any different?

Some months later Undersecretary James Lyons and I were testifying before a Senate subcommittee about land exchanges in Arkansas between the national forests and the Weyerhaeuser Company. Senator Dale Bumpers, Democrat of Arkansas, was in the chair. Two cameras were filming the proceedings.

Senator Stevens stormed into the hearing, late and with fanfare. I imagined that I could hear trumpets and a roll of drums in the background. Almost immediately he instigated a heated exchange with Undersecretary Lyons who, quite adeptly, deflected the attack. Lyons put the ball back in the senator's court. Senator Stevens had come out on the short end—with cameras running. In an attempt to recover, he lashed out at me, but from deep left field. Stevens pointed at the several cameras and cameramen in the room and shouted, "Those cameras are in this hearing room without the committee's permission—without my permission. I want them out of here!"

He pointed at me. "Those are your agency's cameras and you brought them in here without my permission!" He was red, almost purple, in the face and spit bubbles appeared in the corners of his mouth. It wasn't a pretty sight.

After checking with my staff seated behind me, I replied, "Senator, those folks are not USFS personnel."

Now Senator Stevens was on a rip. He snarled, "Well, then, just who do those cameras belong to?

I answered, "Senator, I don't know. But I recognize Patricia Woods, a contractor who facilitates and organizes a considerable amount of training for federal agencies."

The senator lashed back, "And, Chief, just what kind of training would that be?"

With a perfectly straight face, I answered something like the following: "Dr. Woods specializes in training federal and state agency personnel about the interactions of agencies with the Congress. She emphasizes the respectful and balanced give-and-take between the agencies and congressional committees essential to developing mutual understanding and requisite skills to intelligently, rationally, and civilly conduct the people's business."

The gallery and the few reporters present were obviously stifling their amusement. Laughing at the senior senator from Alaska and chairman of the by-god Appropriations Committee was apt to light his notoriously short fuse. Senator Stevens pointed and shook his finger at me. "If your people need to understand all that stuff, why don't you haul them into town and let them see for themselves how it all works?"

I leaned forward and said, in the most calm and reasoned tone I could muster, "Senator, we try to be as efficient with the taxpayers' money as we can. That would just be too expensive. Training films are much cheaper than the costs involved in bringing personnel to Washington, especially all the way from Alaska."

The senator was not mollified. He pointed at the other cameraman and demanded, "Well, whose camera is that?"

I looked at Undersecretary Lyons. He didn't know. I said, "Neither Mr. Lyons nor I have any idea who that camera belongs to. Perhaps the cameraman can tell you?"

Senator Bumpers tapped his gavel to quiet the chuckles. He asked the cameraman, "Who do you work for?" The cameraman didn't answer and seemed to become inordinately interested in his shoeshine. Senator Bumpers tapped his gavel—three times and slowly—and repeated the question.

The cameraman looked up, removed his headphones, and mumbled an answer. The chairman asked him to speak up. The answer came loud and clear. "I work for the Republican National Committee."

Now everybody was laughing—except the cameraman, Senator Stevens, and the senator's aides and witnesses. An aide opened the door at the back of the hearing room as a signal to Senator Stevens that it might be a good time to leave. He took the hint. The aide slammed the door behind him as he followed the senator in a dramatic exit.

I hoped that our subsequent hearings would be less influenced by arrogance and smartass exchanges. Things did get a little better, but not much.

VELCRO

While I served as USFS chief, I sometimes received death threats that were turned over to and evaluated by the agency's law enforcement officers. Most were dismissed as crank threats, and those deemed more genuine were turned over to the FBI. As a result, I was occasionally assigned bodyguards when I traveled.

In the case of air travel, when I arrived at my destination, USFS law enforcement officers assigned to security detail met me at the gate. They identified themselves by holding a small sign with my name on it. One such officer stands out in memory.

Officers assigned to my security detail were typically male, young, tall, and muscular. In this instance, I walked off a commercial flight and saw an attractive young lady in a gray suit with a long wraparound skirt holding up a sign with my name on it. I guessed her to be thirty years old and about five feet seven inches tall; she had an athletic though slender build. She recognized me and informed me that another USFS law enforcement officer would meet us at the curb with a car.

Officers on security detail commonly wear sunglasses to make it difficult for others to discern what they are looking at. They focus on the surroundings and not on the person for whom they provide security. This young lady knew her business. Most USFS personnel are relaxed around the agency brass—including the chief—and conduct themselves informally. That is not commonly so with officers assigned to security detail. They are all business.

As we walked through the airport, I made an effort to tease my protector—just a little—so as to break the ice. When I asked if she was "packing heat," she responded in a flat tone, never even glancing at me as she scanned our surroundings, "Yes, sir!"

Just to show her that I knew one kind of pistol from another, I asked, "Beretta? Nine millimeter?"

She didn't look at me. "I carry a .45 caliber, ACP, Model 1911a—more stopping power, sir!"

We stepped onto the moving sidewalk. She stationed herself between the pedestrians walking alongside and me. It was a funny feeling: a six-foot two-inch, 240-pound male being protected by such a slender, attractive young woman.

I continued, "Can you shoot a .45? As I remember from the military, they have a lot of recoil."

She refused to be distracted. "Yes, sir!"

I asked, in a half-teasing fashion, "Well, how good are you?"

She kept scanning the people around us. "Good enough. I qualify "expert" both on standard targets and combat pistol, sir!"

I kept trying to get a rise out of my protector. "You really think you could shoot somebody?"

She replied, without hesitation, "I pity the poor bastard that wants to find out, sir!"

I detected that she saw nothing remotely amusing about my prattle. I picked up my bags from the carousel and we headed for the waiting car. Standard for a security guard, she remained unencumbered. I looked at her long wraparound skirt. "Can you move fast if we have to?"

She looked over the top of her sunglasses. "Yes, sir!"

"How?"

"Velcro," she replied. With that, she yanked off the skirt, which had been fastened with Velcro at the waist, to reveal running shorts and shoes.

I was convinced that I had a top-notch officer leading my security detail. I became even more impressed as the day wore on. I may be a tad slow sometimes, but I ain't stupid.

I told her that she didn't have to address me as "Sir." She said she would stop calling me "Sir" if I stopped "asking dumb questions, sir!" I complied. But I still smile when I think of Velcro.

After that, anytime a security detail was deemed advisable, my secretary, Sue Addington, asked if "Velcro" was available.

GET OFF THE HORSE AND CLEAR THE TRAIL!

In mid-summer of 1995, I invited Undersecretary Jim Lyons and his young daughter Elizabeth to accompany me on a horseback pack trip into the Eagle Cap Wilderness in northeast Oregon. I was still part owner of a string of horses with my old partner and hunting companion, Bill Brown.

When we arrived at Bill's house in La Grande, Bill had our outfit ready to load and get on the road. All I had to do was purchase the groceries, pack them in the panniers, "manta up" our personal gear into packs, load the gear and horses, and take off for the trailhead. We were at the trailhead and ready to ride out just after noon.

The trail headed downslope through mixed conifer stands—alternating with aspect and elevation between stands of lodgepole, ponderosa pine, white fir, and Douglas fir. Where winds had blown down trees and created openings in the canopy, we could see some of the last flowers of spring. Several miles down the trail we came to an airstrip, an old lodge, cabins, and barns now maintained by the Wallowa-Whitman National Forest as a work center. The whole lash-up had once been an in-holding within the Eagle Cap Wilderness that belonged to outfitter "Red" Higgins. It was still known as Red's Horse Ranch.

A USFS employee saw us riding in across the airstrip and came out to open the gates. She did not recognize me as chief of the USFS (I was not in uniform) nor did she recognize Undersecretary Lyons. Bill, whom she did recognize, was riding in the lead and introduced us only by first names. He asked if the trail was open up to one of the upper meadows where we intended to camp on the way to Blue Lake, a high mountain cirque lake perched just below tree line. We were informed that some logs were down across the trail but that with a little effort it was possible to ride all the way.

We headed out for a "hidden" campsite where Bill and I had often camped in the past, especially when hunting for big mule deer bucks. The fishing for small brook trout in the adjacent stream was a little tricky but could be a profitable enterprise. From the hidden camp it was but a couple hours' ride on up to Blue Lake, where the fishing was usually what Bill described as "primo."

After we made camp, I showed Jim and Elizabeth how to roll-cast with a fly rod. We knelt and roll-casted under the old hemlock trees and ocean spray shrubs—a few still in bloom—into fishing water. The moss on the rocks and banks reflected the cool dampness so different from the dryness and heat just a few yards upslope. This was a new way of fishing for Jim, so we spent considerable time retrieving flies from trees and bushes. In spite of that, we caught enough trout for a fine supper and released several more.

The next morning, after a breakfast of trout that we kept chilled overnight in the stream, we were off for Blue Lake. We rode several miles up the trail before we began passing through stands of lodgepole pine. Lodgepole pines are shallow-rooted and grow in dense stands, which allows them collectively to withstand wind. Where lodgepole stands are "opened up," such as along a cleared trail, adjacent trees are much more apt to blow down.

Sure enough, we soon encountered trees down across the trail. Previous riders had simply ridden around the downed trees, tearing up the ground cover and leaving scars susceptible to erosion. The downed trees were of a size commonly referred to as "pecker poles," ranging, in this case, from ten to fifteen inches in diameter breast high. Both Bill and I were carrying double-bitted axes in scabbards. Where the trail was blocked by a pecker pole or several, we dismounted and cut the obstruction out of the trail. Often all we had to do was cut the trunks in two and then pivot the lighter end of the tree to clear the trail. Jim, a forester by training—at Yale no less—proved reasonably adept at wielding an ax. In the course of the day I guessed that we cut some twenty-five to thirty pecker poles out of the trail.

Elizabeth Lyons with Jack on the North Minam River, Eagle Cap Wilderness, August 1996. The daughter of Undersecretary of Agriculture Jim Lyons, Elizabeth proved to be a quick learner with the fly rod.

Jim finally asked me, "The lady down at Red's Horse Ranch obviously knew that USFS folks were routinely riding up and down this trail. Why didn't they clear the trail?"

It was a good question—a very good question. I was embarrassed as I answered. "I really don't know. Maybe they didn't have an ax with them."

Jim responded, logically enough, "Why not? You guys are packing axes." Having made his point, he let the matter drop.

We rode on up to Blue Lake, where we had a wonderful day of fishing, dined well on trout—which I fried up on the spot—and worked in a nap in the afternoon sun. We arrived back at camp just after dark. Elk steaks on the grill were the perfect end to a fine day.

There was no need to pitch tents, as the sky was crystal clear and there was not a breath of wind—and, truthfully, we had checked the long-range weather report before we departed. There was a canopy of stars such as simply cannot be seen from "down below where the people is."

After a week, moving from camp to camp, we came again to Red's Horse Ranch on our way out to the trailhead. Two USFS employees, one of them the woman we had met on our way in, greeted us. She asked, "Were you able to ride up to Blue Lake? Those pecker poles across the trail can be a real problem."

"Not any more," I replied. "The chief of the USFS and the undersecretary of agriculture—with some help from our seventy-five-year-old friend Bill—cleared the trail, except for a couple of trees that were just too big for us." Now both USFS folks realized who we were. "Have USFS folks been riding around the downed trees and screwing up the trail in the process?"

"Well, Chief, our budget for trail crews got cut back this year. The trail crew didn't get that far before we ran out of money."

"That doesn't answer my question. USFS employees rode around those pecker poles—likely several times. Why not just take the axes out from under your legs and clear the trail?"

"We didn't have axes. We're not the trail crew."

Now I wasn't smiling. "I suggest that *anytime* you folks are riding in the wilderness, you carry an ax under your leg. And it would be a really good thing if, instead of riding around a pecker pole across the trail, you got down off your horse, took out your ax, and cleared the trail. If Undersecretary Lyons and Chief Thomas and their seventy-five-year-old traveling partner can clear out pecker poles, I suspect you can, too. Okay?"

Both nodded. I smiled, but not an overly friendly smile.

We visited a while longer about other things, just to make sure we parted on collegial terms. After we rearranged our manta packs, cinched up the riding saddles and pack saddles, and stepped up onto our horses, I looked down at my colleagues. "Folks, I'll be back in this country in a couple of months for a little elk hunting with my partner Bill here. I really hope I don't have to cut any pecker poles out of the trail." I smiled. "It's been good visiting with you. Keep the faith!"

Bill, leading three packhorses, led out across the airstrip headed for the trailhead. I followed, keeping an eye on the pack string, while Jim and Elizabeth took up the drag. I turned, took off my greasy old Stetson, and waved good-bye as we moved into the trees.

As promised, a couple of months later, I was back in northeast Oregon to join Bill for our annual elk hunt. Snow was on the ground. We traveled over fifty miles on trails in the Eagle Cap Wilderness during our ten-day hunt and didn't have to cut a single pecker pole out of the trail.

When I got back to Washington, I sent the supervisor of the Wallowa-Whitman National Forest a handwritten note commending him and his staff on the condition of the trails in the Eagle Cap Wilderness. He wrote back to tell me that it had become *de rigueur* for USFS folks traveling horseback in the Eagle Cap Wilderness to carry an ax—and to use it when they came to a pecker pole across the trail.

There was, I thought, a lesson here. In life, when you come to a "pecker pole" across the trail, don't ride around. Step down off your horse, untie your ax, and clear the trail. It makes life better for everybody. After all, there can't always be a trail crew to do the dirty work.

RUBBING IT OFF ON THE OTHER BOOT

In the summer of 1992, the Foothills Fire swept across the Boise National Forest in Idaho, burning over some 257,000 acres between Boise and Mountain Home. My longtime colleague and fellow "old-time" wildlife biologist, Stephen "Steve" Mealey, was the forest supervisor. He was an aggressive and confident man by nature, a can-do sort from the old school. Even before the fires were out, Mealey and his team was planning for post-fire timber salvage and forest restoration.

Within a month after dropping temperatures and fall rains gave the fire crews the edge they needed to extinguish the fires, Mealey and regional forester Gray Reynolds jointly determined to carry out one of the most extensive salvage and forest restoration operations in recent USFS history—including salvage of an estimated 140 million board-feet of fire-killed timber. They exercised their authority to suspend the administrative appeals processes, avoiding what they anticipated would be significant, expensive delays in the salvage efforts.

They believed that such delays would significantly reduce the quality and market value of the dead timber, thereby reducing the economic feasibility of extensive salvage efforts. That is, if they moved fast, an extensive salvage operation made good economic sense. The first restoration operations were completed, essentially without a significant hitch, late in the fall of 1992.

At that point, Reynolds was promoted to be deputy chief for the National Forest System at the USFS's headquarters in Washington, D.C. He was replaced as regional forester by Dale Bosworth (who, within a decade, would become chief of the USFS). Both Reynolds and Bosworth supported Mealey's plans when they were presented to me.

When I became chief in late 1993, criticism from some in the environmental community over the salvage operations on the Boise National Forest was building to a crescendo. Mealey and his key staff came to Washington to brief the USFS brass and the secretary and undersecretaries of agriculture. We were informed, in detail, about what had already taken place and future plans for salvage of merchantable dead and dying trees. Thinning of trees from overcrowded stands that had not been touched by the fires were planned to reduce the likelihood of additional stand-replacement fires.

The briefing was exceptionally well done, reflecting Mealey's experience as a U.S. Air Force briefing officer on General Westmoreland's staff during the Vietnam War. Some said we should have filmed his performance for use in training sessions.

I asked, "What do you mean when you say that your team 'moved aggressively' in salvage operations?"

Mealey replied, "When a district ranger asked me how quickly we should proceed, I told him 'if the logs aren't still smoking when they reach the mill, you're lagging behind.'" Hyperbole? Yes, but he was leading out when others were simply wringing their hands.

Mealey, Reynolds, and Bosworth were old hands in the management of national forests—honorable, skilled professionals with *chutzpah*. They were second-generation USFS employees and were said by some to "bleed green" (the USFS colors). As I trusted their collective judgments and motivations, I instructed them—Mealey in particular—to proceed on course. I promised him that Reynolds, Bosworth, and I would have his back.

Undersecretary James Lyons initially concurred with our decision, and I naïvely assumed that Lyons had the political appointees above his pay grade fully informed and on board. By the end of 1994, timber salvage operations were completed, with what the USFS considered appropriate environmental constraints.

The salvage operation had been carried out in record time and, as a result, had provided the best possible economic return to the federal treasury. Mealey's team was equally expeditious with restoration work, such as planting seedlings and completing efforts related to soil stabilization. They simultaneously moved forward with plans to thin, over the next decade, thousands of acres of overstocked stands to lessen chances of even more stand-replacement fires.

Environmental organizations, including the Idaho Wilderness Society and the Alliance for the Wild Rockies, were caught flat-footed when Mealey, with approval from up the chain of command, suspended standard appeals processes, which he had the authority to do. They considered that both inappropriate and a personal insult. Critics, caught off-guard, were angered when the required environmental assessments, reviews, and other procedures were satisfactorily accomplished much faster than the norm.

That speed thwarted the use of the standard operating procedures of some environmental organizations aimed at delaying, or negating, almost any proposed management actions that involved the salvage of fire-damaged timber. In addition, spokespersons for some environmental organizations loudly opposed the plans for forest thinning as a Trojan horse. Many of them considered the overall operation as "just another example of the USFS colluding with the timber industry to 'get out the cut.'"

As one effort after another by environmental organizations failed to derail or delay operations, frustrations mounted. Mealey's larger-than-life persona, coupled with his style of leading from the front and his willingness to accept full responsibility for his decisions, made him a perfect target. He was vilified by some as "the Butcher of the Boise." Backed up by the USFS's regional forester, the chief, and the undersecretary of agriculture, he did not significantly alter either his stance or approach.

In the meantime, some hard-core environmentalists, with some success, were cashing in the political chips garnered in the last presidential election. Clinton administration political appointees, who had initially pushed the USFS hard and persistently for aggressive salvage operations, quickly withered in the building political heat. As criticism mounted, the politicians and political appointees moistened their forefingers and held them high to discern the direction and force of the political winds.

Staff morale on the Boise National Forest, which had soared with recognition of their outstanding performance, began to wilt. Bosworth, with support from Reynolds, saw to it that Mealey and his top staff were recognized for outstanding performance. Mealey, in turn, saw to it that deserving subordinates were appropriately recognized.

Some months later, the Senate Committee on Natural Resources and the Environment held field hearings around the West concerning the just-completed, pending, and ongoing timber salvage operations. In the late summer of 1994, I was called to testify at the field hearing in Boise.

I arrived several days before the hearing so that Mealey and his staff could expose me to both completed and ongoing salvage and restoration operations—from the air and on the ground. I saw it all firsthand, "the good, the bad, and the ugly." I concluded that, with some rare exceptions (which should be expected in any large-scale salvage operation carried out under tight timelines), the USFS could be proud. I was able to stand behind Mealey and his team with confidence.

On the day of the hearing at the Boise City Hall, I made my way through the crowd toward the entrance. Protesters blocked the entrance to the building. I was accompanied by Mealey, Bosworth, and several USFS law enforcement officers, all of us in uniform. The officers carried side arms. As we worked our way through the crowd, my path was blocked by a protester, yelling obscenities and waving a sign that disparaged both the ancestry and intellect of "the Butcher of the Boise" and also castigated the USFS.

I constrained myself from punching him in the nose and setting him down unceremoniously on his fat ass. Then he spat in my face and whacked me on the head with his sign. I started for him with full intent to physically impress upon him the error of his ways. Fortunately, the officers intervened. One grabbed me by the belt and another cleared the protester out of my path. By the time I was seated on the front row in the hearing room, my blood pressure was slowly receding toward normal. A minor-league bloody nose and a sore head added to my irritation.

Senators Larry Craig (Republican from Idaho) and Tom Daschle (Democrat from South Dakota) conducted the hearing. After their initial remarks, I was the first witness. In my opening statement, I said that the staff of the Boise National Forest had been commended for their performance to date. Senator Craig asked my opinion specifically about the performance of forest supervisor Stephen P. Mealey.

In the days before the hearing, Idaho's governor and congressional delegation told Secretary of Agriculture Mike Espy and me that they thought Mealey and his staff had performed in an exemplary fashion. Senator Craig's questions were what experienced observers of government hearings refer to as "batting practice pitches"—slow, straight, right down the middle, and easy to hit out of the park. When really angry, I unfortunately tended to say exactly what I was thinking without sending it through the political correctness filters that top-level bureaucrats develop, or should develop, between their brains and their mouths.

I replied, "Senator Craig, I told Mr. Mealey to move aggressively with the salvage and thinning operations that he proposed and that had been approved by the regional forester. I gave that same instruction to several other forest supervisors dealing with post-fire situations. While some stood around wringing their hands, Mr. Mealey carried out his instructions. He and his staff did, and are doing, exactly what they were approved and instructed to do—with dispatch. I won't back away from them now. If Supervisor Mealey was wrong, the regional forester was wrong, the deputy chief for the National Forest System was wrong, and I was wrong."

The two senators had their hands over their mouths, but their eyes gave their smiles away. My language may have been both impulsive and somewhat politically incorrect, but my point was clearly made.

The press grabbed onto the colorful—if somewhat intemperate—statement. Political overseers in Washington flinched. But rebukes and cautions were mild, just enough to mollify those who protested my vulgarity. But at least in this case I was accurately quoted.

STRIPPING FILM AT STORM KING

In the middle of the night of July 6, 1994, the phone ringing beside my bed roused me from a deep sleep. In fumbling for the phone, I knocked it to the floor. When I finally answered, I recognized the voice of Lamar Beasley, USFS deputy chief for administration.

His voice cracked. "Chief, I have terrible news." A long silence. "We may have lost as many as forty firefighters in a blowup of a wildfire on Storm King Mountain just above Glenwood Springs, Colorado. It's certain that at least ten firefighters are dead. At least thirty more were up on the mountain and are unaccounted for."

I sat up, wide awake but stunned into silence.

"Chief? Chief? Are you there?"

"Yes, I'm here. I'm struggling to come to grips with what you said." After a long pause, "I know that there is a standard operating procedure for a situation like this. Who activates that procedure? Specifically, what do I need to do?"

I was informed that a rotating roster of USFS senior executives exists to be mobilized to institute, organize, and lead investigations.

The deputy chief for legislative affairs, Mark Reimers, was at the head of that duty roster and was already en route to the Washington office to meet other key staff to organize the investigation.

The firefighting operations at Storm King Mountain were under BLM management. However, as the known fatalities were USFS firefighters, that agency would lead the investigation. Team members from across the country were already on the way to Grand Junction via civilian, government, and chartered aircraft.

Clearly, Beasley had the initial critical response activities firmly in hand. I asked for the home phone number of the BLM's acting director, Dr. Michael "Mike" Dombeck. I also asked Beasley to coordinate with the BLM to get Dombeck and me on the first available flight to Denver and arrange for a USFS plane to fly us on to Glenwood Springs.

Both Dombeck and I had been on our cell phones almost constantly when we met at Dulles International Airport and compared notes. We now knew that twelve firefighters—including USFS smokejumpers and a Hotshot crew—were confirmed dead. Two members of a BLM "helitack crew" (firefighters put on the scene via helicopter) were missing.

Only one of the group caught in the blowup, smokejumper Eric Hipke, had survived. He was badly burned but expected to live. All the other firefighters initially declared missing had walked out onto the highway at the foot of Storm King Mountain—exhausted, frightened, but alive and physically unscathed.

Dombeck and I were met at the Denver airport by deputy regional forester James Lawrence, an old friend from his days as supervisor of the Umatilla National Forest, just over the hill from La Grande. He escorted us to a waiting chartered aircraft. On the short flight to Glenwood Springs, Jim told us the on-the-ground investigators had instructed that, until they could survey the scene, the bodies of the casualties would be left in place. It seemed heartless, but it was the technically correct thing to do.

Upon landing, Dombeck and I were whisked away to a meeting with Secretary of the Interior Bruce Babbitt and Colorado's governor, Roy Roemer. On the way, we saw dozens of hand-lettered signs held by citizens of Grand Junction—"Thank you, firefighters," "God Bless the USFS," "We are so sorry!"—and many more. I was struck by how circumstances of the moment can change attitudes. I could not help but recall a Kipling poem that I had memorized for a high school English class: "For it's Tommy this, an' Tommy that, an' 'Chuck him out, the brute!' / But it's 'Savior of 'is country' when the guns begin to shoot."

The meeting quickly focused on assuring the press—and through them, the public—that a thorough, transparent, full-scale investigation was under way. The press, especially the television crews, were clamoring for information, photos, and interviews. Some reporters questioned the government's motives for keeping the press out of the area where the firefighters' bodies still lay. A coordinated effort—approved by Dombeck and me—was made to prevent the press from photographing the bodies. We owed the families of those who had died that simple consideration. But some reporters were on the trail of just what might be a bigger and better story. Their competition seemed to be building to a feeding frenzy.

One photographer managed to disguise himself as a firefighter by somehow obtaining a dirty yellow fire-resistant Nomex shirt and green pants, smokejumper boots, and a battered hard hat labeled "USFS." He was initially assumed to be a firefighter or perhaps a member of the investigative team. His scam worked until a USFS law enforcement officer noticed that his face lacked both tan and grime and his hands had clean, manicured fingernails. This guy, whoever he was, was no wildland firefighter.

When the officer asked to see his "red card," the identification of a *bona fide* wildland firefighter, the impostor admitted to being employed by a newspaper while vociferously maintaining his "rights" to take photos of the bodies. The officer took him into custody and confiscated his camera and film.

Now what? This was as much a political call as it was a violation of law. The buck quickly ended up in my lap, where it belonged. I was talking with dozens of members of the press when I got the word. One shouted, "Do you have your cover-up under way?" I had never felt so disgusted or angry with the Fourth Estate as at that moment.

The press demanded answers we did not have. Some repeatedly demanded to know, basically, "who screwed up?" To be fair, I could see that it was a huge story. I knew that it was the reporters' job to garner and feed information to newspapers, radio, and television.

Finally, in exasperation, I stole a line from a man I much admired—Harry Truman. After all, I was responsible for the USFS and the lives of *my* firefighters (though the BLM was in command for this fire). So I told the press, "We don't know exactly what happened—not yet. Nobody does. We are doing everything we can to find out. When we know, you will know. But in the meantime, the buck stops here—with me." That seemed to calm the feeding frenzy, at least for the moment.

I was frazzled, dead tired, and sick to my stomach. Someone guided me into an empty room and shut the door. A USFS law enforcement officer, with the errant photographer in custody, brought me up to speed. The photographer identified himself as a photojournalist and identified his employer. When I asked him where he got the firefighter's outfit, he smiled and said that was a "trade secret." He demanded that I return his camera and release him.

I asked, "My God, how could you take pictures of the bodies of those young people for publication? Their parents, spouses, brothers and sisters, and friends would be haunted by those photos for the rest of their lives. Surely they deserve more respect and consideration than that!"

The photographer just shrugged his shoulders and said, with a cynical half-smile, "That's show business."

I flushed and gritted my teeth. My anger quickly morphed into disgust, and I asked for the camera and told the officer to leave the room. I opened the camera and exposed the film, thereby destroying whatever images were present. I handed the photographer his camera and bag and called the officer back into the room. I told him to escort the photographer out of the secured area and, figuratively, boot his ass down the road. Fortunately, the officer was an old hand and knew which instructions to follow and which ones to ignore. He escorted the photographer out of the secured area and released him—and likely mentioned that felony charges for impersonating a federal official and various other violations were under consideration.

I called Undersecretary Lyons in Washington to give him a heads up. Bless his heart, he said, "Don't worry. You did right. I've got your back." Jim stood tall when it counted. I never heard any more about the photographer. Sometimes what's right may not be exactly legal—and vice versa. I thought then, and think now, that what I did that day was the right thing to do. It had been a long and miserable couple of days, the longest and worst days of my professional life.

What had been a relatively benign fire in rugged terrain on Storm King Mountain simply blew up and killed fourteen elite wildland firefighters. It was clear from the get-go that there were problems in dealing with that wildfire appropriately and safely. This was not the first such disaster and surely would not be the last. But this one had occurred on my watch, and the buck stopped with me—and BLM director Mike Dombeck.

Reviews of the situation by a team appointed by Dombeck and me—and a separate review by the Office of Safety and Health Administration—came to similar conclusions save for one. That separate report indicated that the supervisory firefighters were both careless and essentially uncaring.

I objected vehemently to that addendum. I did not, and do not, think that was true in any sense. I cared—deeply. The fire bosses all took the losses personally. In my opinion, it was either unthinking or a crass, politically motivated "cover your ass" statement—or both. Maybe I cared too much and was much too sensitive, but I haven't changed my mind.

Changes in firefighting protocols were immediately instituted (and others have been made since) aimed at significantly reducing risks to wildland firefighters. Those increased precautions may have reduced the effectiveness of fire suppression efforts, but they were necessary. The price in human lives paid during the 1994 wildfire season was a wake-up call, one that came just as recognition of the potential for more catastrophic wildfires was growing.

Decades of effective wildfire suppression, dramatic reductions in active forest management, increasing tree mortality from disease and insect outbreaks, probable global warming, and an ongoing lack of consistent policy direction from successive administrations and Congresses combined to create a thickening witch's brew for more fires and hotter, bigger fires. Equipment and personnel costs were increasing while pressures to cut budgets were increasing. It was a scary scenario then—and one that became more frightening with each passing year.

Clearly, the lesson that emerged from the wildfires in and around Yellowstone National Park in 1988 coupled with Storm King Mountain fire in 1994 has yet to produce a sufficiently clear, coherent policy—and the necessary funding—for federal and state land management agencies to deal with wildfires. Rocky Barker, in his book *Scorched Earth: How the Fires of Yellowstone Changed America*, discussed these issues and told parts of this story.

Neither his nor my questions and concerns have been satisfactorily addressed. How many more political, economic, human disasters will be required before we blaze a new trail to deal with a worsening situation relative to wildfires? If global climate change is real, and the vast majority of scientists think it is—including me—the situation will only get worse. Time's a-wastin'.

HIPKE'S TICKET HOME

One of the first duties Mike Dombeck and I addressed in Glenwood Springs was meeting with the firefighters who had taken different routes down from the hell on Storm King Mountain and escaped unscathed, at least physically. Our next duty was to meet with the parents of two firefighters who had been dropped off by a helicopter and were still missing.

They asked us if there wasn't someplace—creek, lake, pond, cave, rocky outcrop—where the firefighters might have survived the inferno and were awaiting rescue. In our hearts both Mike and I knew that the young firefighters were lying dead somewhere up on Storm King Mountain. We looked at each other, sorely tempted to lie to keep hope alive—if just for a little while longer—but we couldn't bring ourselves to do that.

I listened with tears welling in my eyes as Mike—these were his firefighters—told them there was no realistic hope for their survival.

We assured them that everything possible was being done to locate their sons. Mike's and my terrible distress paled to insignificance as we looked into their eyes.

We took our leave and went directly to the hospital to visit Eric Hipke. Hipke, a smokejumper from Washington State, and his fellow firefighters had run for their lives up the steep fire line toward the ridge as the fire exploded behind them. Some combination of superior physical conditioning, stamina, fear, the will to live, and luck propelled him past the others scrambling toward the ridgeline. Just as he reached the ridge, he was literally blown over by the wall of flame exploding up the mountain. He was badly burned over his back, neck, and arms, and hands, from which skin was hanging. But he, and he alone, was very much alive.

Mike and I took the back stairs to avoid the gaggle of increasingly aggressive reporters lying in ambush in the lobby. When we reached Eric's hospital room, we were pleased to find his parents there. Eric was in remarkably good spirits due, I suspected, to a combination of good drugs and the euphoria that comes with having escaped almost certain death. We had no intention of discussing the previous day's events. Trained investigators would take care of that at a more opportune time. We simply wanted to know if we could do anything for Eric and his parents.

Eric's parents asked if we could get him to the burn center in Seattle, one of the best in the world and in their hometown. We in turn asked the doctors whether that was medically advisable, prudent, and feasible. The answer was yes to all questions. However, Eric would have to be moved via a specially equipped Medivac jet. Though that wouldn't come cheap, there was no discussion of the costs. The hospital's administrator offered to provide, without charge, appropriate medical personnel to accompany Eric on his journey. It was an unsolicited gift to Eric from the people of Glenwood Springs.

I looked to Dombeck. He nodded yes. I turned to a USFS administrative officer, who was standing by to see that Hipke got everything he might possibly need. He took me aside and whispered, "Chief, you can't authorize such expenditures. There is no clear reason he needs to be airlifted to Seattle. The bean counters and political critics are likely to view this as an unnecessary, even extravagant expense."

He was doing his job, and we suspected that he was, at least technically, correct. I asked for his clipboard and wrote something like, "I hereby authorize airlifting Eric Hipke and his parents to Seattle by whatever means are deemed appropriate by the medical staff in Glenwood Springs. I assume full responsibility for this decision." I signed it with the date and time. Dombeck signed on. I handed the clipboard back and said, "Make it happen—ASAP."

The administrative officer said, "Chief, I gotta tell you one last time: you guys may end up paying for this out of pocket."

We stepped out into the hallway for privacy. I told him, "I understand what you are telling me, and I know you're just doing your job. We appreciate that. However, we don't think our decision will be questioned—that would be just too politically incorrect. If we do end up with the bill, I have every confidence that our employees—and maybe the people of Glenwood Springs—will help us cover the tab."

On the way down in the elevator I had never felt so exhausted and so wrung out emotionally. I literally felt my knees sagging. I straightened up and told Dombeck that I was reminded of a line from a Cecil B. DeMille film in which Pharaoh summarized his decisions by saying, "So let it be written, so let it be done!" I wasn't nearly as confident as all that, but it was written and it would be done.

Our decision was not questioned.

RED CARD QUALIFICATIONS

All USFS personnel who are authorized to engage in firefighting activities are issued a "red card" that attests to their identity, status, qualifications, and restrictions on activities related to firefighting duties. Employees in the agency's research division are ordinarily excused from firefighting assignments. We researchers were sometimes laughingly told that we were last in line for firefighting duties right behind seriously wounded war veterans and near-term pregnant women. As one veteran firefighter once said to me, "Things ain't likely to ever be that dire." So I in all my twenty-eight years in the USFS, I had never had a red card.

The 1994 fire season, which took place my first summer as chief, was deemed among the most severe since the "Big Burn" of 1910. There had been numerous fatalities and severe injuries among wildland fire-fighters. I spent much time during that fire season "showing the USFS flag" by visiting fire camps across the West to demonstrate to those on the fires lines that we knew they were there, were concerned for their welfare, and were appreciative for their service. Toward the end of fire season, federal land management agencies were shaking out every employee with a red card—and those who could have their red cards renewed—for assignment to firefighting operations.

To symbolize my concern for and solidarity with the men and women on the fire lines, I told my secretary, Sue Addington, to ar-range to have me issued a red card. Before the end of the day, the di-rector of fire and aviation management came to my office, along with several of her staff for a brief ceremony—including a bit of welcome levity—before handing me my very own red card. I only glanced at the card before putting it in my wallet.

Less than a week later, I was visiting a fire camp on the east slope of the Cascade Mountains in Oregon. After the camp quieted down for the night, I joined the Incident Command Team for the last cup of coffee of the day. The incident commander stood and tapped a spoon against a cup to gain attention. He turned to me. "Chief, I hate to be a stickler for details. But you have made it clear that we will be in strict compliance with all applicable orders, rules, and regula-tions when engaged in firefighting activities. It is clearly stated that all USFS personnel, which I would assume includes you, must possess a red card to be engaged in firefighting activities of any kind, including observations thereof. So, Chief, I must ask to see your red card."

I stood up and replied, "Well, hell yeah, wiseass, I've got a red card. You really want to see it?" It was a challenge.

The commander wasn't smiling. He held out his hand. "Yes, sir! Please."

I fumbled through the pockets of my Nomex shirt and jacket to find my wallet, sorted out my newly issued red card, and handed it over with a flourish—and a wink at the crowd. The commander took the card, made a show of retrieving his reading glasses from the pocket of his Nomex shirt, and perching them on the end of his nose. He held the card at arm's length and inspected it, front and back, with exaggerated care. He handed my red card to the deputy incident commander, who repeated the performance. My red card was passed around until the entire team had carefully perused it.

The incident commander handed the card back to me. "Chief, please read the restrictions listed on the back aloud so that all concerned understand your capabilities and limitations related to firefighting duties."

Under "Restrictions" the card read (in capital letters, no less), "DO NOT LET OUT OF THE SIGHT OF QUALIFIED FIREFIGHTERS—NO SHARP TOOLS!" The tent exploded in laughter. It was an honor to work for an outfit in which the troops could yank the boss's chain—in appropriate circumstances—and get away with it. This was such a time. I was honored.

BUT I HAVE TO FLY IN THAT PLANE

I made it clear during the wildfire season of 1994 that I was not an experienced wildland firefighter. I made no pretense of taking charge of any fire-suppression operations or even offering suggestions.

The last three of the twenty-four fatalities that marked that seemingly never-ending wildfire season resulted from a helicopter crash in New Mexico in mid-July. One firefighter survived the crash in which two pilots and a firefighter perished. I was visiting a nearby firefighting operation when the tragedy took place and was at the helicopter base early the next morning.

The father of the dead firefighter was himself a longtime USFS firefighter who, oddly enough, was in charge of helicopter operations for the fires in question. In spite of his tragic loss, he accompanied me as I paid my condolences to the families of the other young men killed and then visited the lone survivor in the local hospital.

I moved on to visit other firefighting operations, having no idea if my visits were doing any good, but that seemed better than sitting in my office in Washington dreading to hear the phone ring. I had become increasingly saddened and frustrated as the wildfire season wore on and the fatalities and injuries mounted.

In early September, I attended a memorial service in New Mexico for the three people who had perished in the helicopter crash. Joining me at the ceremony were Secretary of the Interior Bruce Babbitt, Secretary of Agriculture Mike Espy, and Mike Dombeck, acting director of the BLM. It was time to "show the flag" to emphasize the agency's and the nation's gratitude to wildland firefighters and their families. The memorial service was held in a public park. The speaker's platform was backed by a band shell. I was nearly played out—mentally, physically, and emotionally—and don't remember much of what was said. However, it was clear that those in attendance appreciated the two cabinet secretaries and the two agency heads being present.

I was the last to speak. The eloquent, heartfelt comments of the clergy and the other officials had left me little to say. Just as I finished my brief remarks, lightning flashed, followed seconds later by booming claps of thunder. Within minutes, a thunderstorm, so long beseeched of Providence to douse the seemingly never-ending fires of the summer of 1994, swept over us. The rains had come! Hallelujah! Hallelujah!

After the ceremony, we were driven to the airport. The two secretaries had flown in on a twin-engine military jet designated for transporting VIPs. I was standing with them as they waited on the tarmac. A young, unshaven man in a wrinkled uniform walked up to us from where he had been standing next to a single-engine Cessna with the USFS insignia. I recognized him, having flown with him on several occasions over the past decade. I introduced him to Secretaries Espy and Babbitt.

The pilot picked up my duffel bag, threw it over his shoulder, and started back toward the Cessna. I turned to bid the secretaries farewell when Secretary Espy asked, "You don't intend to fly out of here in that plane, do you?"

"Yes, sir, that's my ride."

"You're an agency head. You should be flying only in twin-engine or bigger aircraft."

I smiled and replied, "Mr. Secretary, I am the chief of the Forest Service, not the secretary of agriculture."

He seemed to interpret my response as either smartassed or flippant, though I didn't mean it that way. So I amplified my answer. "Mr. Secretary, I appreciate your concern. But our folks watch what their leaders in the Forest Service and the BLM do. So if we want their respect, we can't ask for or accept any margin of safety or comfort not available to all our employees. Please understand." I turned and followed the pilot to the Cessna.

Once we were seated and strapped in, the pilot looked over at me. "Jeez, Chief, we could have found a twin-engine aircraft for you if we had known there was a problem."

I looked at him out of the corner of my eye and asked, "Are you telling me that this aircraft and perhaps its pilot are less than safe?"

He smiled and nodded. "Seatbelt fastened? We'll just do the best we can with one engine and one old smelly tired pilot." He revved up the engine, taxied out, and gave a fair imitation of Jackie Gleason's famous line, "And away we go!"

We were off to another fire camp somewhere. I trusted the pilot to know where we were going, as I didn't. Right then I didn't care. I was dead asleep before we reached cruising altitude. It had been one more very long day toward the end of a long and deadly fire season. And it wasn't over yet. Secretary Espy, upon reflection, must have seen my point or perhaps chose not to make an issue of it. Or maybe he just forgot about it. Anyway, I never heard any more about it, and I continued flying in single-engine aircraft when circumstances made that a logical choice.

THREE STARS IN THE WINDOW

Late in the very long wildfire season of 1994, I visited a fire camp in northern Idaho. By this time, rosters of all federal, state, and local civilian personnel who were trained and qualified to fight wildfires were tapped out. All available contract crews were on the fire lines. National Guard resources were committed. As the fire season inexorably dragged on into the fall, assistance was requested from the regular military, a very unusual move. Without hesitation, the Pentagon assigned one battalion each from the Army and the Marine Corps to assist in dealing with wildfires in Idaho. By the time I was able to visit their operations, the fires had been contained and mop-up operations were under way. The battalions were packing up to move out.

I told the USFS liaison officer with the military contingents that I wanted to visit to say "Thanks, good job, and Godspeed." Within an hour, a Marine helicopter arrived to pick me up. This was no military surplus Huey contracted for firefighting duties. The ship appeared relatively new and clean and shiny and had a red sign with three gold stars in the window, the "flag" of a lieutenant general. No sooner had the skids of the chopper settled in the dust of the helipad than a Marine Corps major and an Army major stepped out, followed by a Marine Corps sergeant major. They were all spit and polish in their fresh-pressed, starched fatigues. They greeted me with snappy salutes and extended their hands in greeting.

I was informed that the sergeant major would accompany me. He was a handsome man with an appearance and bearing straight out of central casting: close-cropped white hair, gleaming white teeth, and a deep baritone voice—six feet three inches of pure lean and mean.

Once airborne, assuming I was being loaned a general's helicopter, I asked him about the red sign with the gold stars. The sergeant major informed me that the stars represented the equivalent military rank of my position in the civilian government hierarchy. I got a kick out of that. My highest military rank in the U.S. Air Force had been lieutenant. I had no idea that there were such equivalents—or that anybody would know or care.

In spite of being dirty, tired, and frazzled when they picked me up, I was perking up considering my "ride"—a shiny helicopter, a general's flag in the window, and a Marine Corps sergeant major as an escort. At each stop I was greeted by a company commander, a captain, who escorted me to thank the assembled troops. The sergeant major was always exactly two steps to my left rear.

After our last stop, a driver in a safety-green jeep with USFS decals was waiting to pick me up. The sergeant major and I waited for the rotors to wind down before we stepped out of the helicopter. When we were clear of the still-turning blades, I thanked the sergeant major and waved my thanks to the pilots. The sergeant major handed me the window flag with the general's stars, stepped back one step, and snapped a salute. I came to attention and returned the salute. Old habits are hard to break, and it was clear that we shared mutual respect.

Then he leaned forward to yell above the sound of the slowly turning rotor blades. "Sir, you need a haircut and you could use a shave—rather badly." He stepped back, saluted one last time, winked, and stepped up into the helicopter. And they were off in the proverbial cloud of dust.

When I turned around, my driver was leaning back against the fender of a filthy jeep with his hands thrust deep into his pockets. He was dirty, unshaven, and, quite obviously, way beyond tired. He looked me up and down. As we bounced down the bulldozed fire line, he said with a sparkle in his eyes, "You know, I've worked for the Forest Service for over thirty years, and you're the first chief I've ever seen up close and personal. I don't know what I expected, but I sure didn't expect you to be just as dirty and smelly as the rest of us."

We laughed. We didn't talk much on the way back to the staging area. It had been a long, disastrous, deadly fire season. And it wasn't over.

After chow, my driver drove me out to a dirt airstrip where a USFS aircraft, with one engine and one pilot, waited to take me to another fire somewhere. The fires and fire camps were beginning to run together in my mind. I was asleep before we cleared the end of the runway. I didn't put the red sign with the gold stars in the window.

But I still have that flag stashed away somewhere.

C-130S AND THE BATTLE OF LONG ISLAND

In the fall of 1996, six USFS retirees, including me, sat around a campfire in our elk-hunting camp in the high Wallowa Mountains in northeast Oregon, sipping camp coffee. As the fire burned down to coals, the conversation turned to tales of the biggest foul-ups we had been associated with in our careers. I went last.

During the summer of 1995, I was entangled in what some USFS insiders had come to refer to as "the Battle of Long Island." I was in Alaska reviewing several ongoing controversial timber sales when a chartered floatplane dropped me off in Sitka. A message was waiting from the White House operator (never good news) directing me to call Harold Ickes. For a moment, I thought I was caught in a time warp. The only Harold Ickes I could recall had been secretary of the interior under President Franklin D. Roosevelt.

It was about 5:00 p.m. in Alaska, 10:00 P.M. in Washington, D.C. The message said "urgent." So I dialed the number and told the operator who I was and that I had been instructed to return an "urgent" call to Harold Ickes. I wheedled out of her that the extant Mr. Ickes was an assistant to President Clinton. Once on the line, Mr. Ickes came directly to the point, none too gently: "What are you going to do about the fires on Long Island?"

I had been flying over the islands of Alaska for two days and thought he meant one of those islands. "Sir, it has been raining here for the last couple of weeks; there are no fires in Alaska, at least none that I know of."

With some exasperation, he replied, "No, I mean the fires on Long Island, New York!" He obviously had no idea that I was in backcountry Alaska.

I replied, "I don't believe the Forest Service has any responsibility for fighting fires in New York State. And if there is any involvement, I am certain that the acting chief in Washington is fully informed, engaged, and authorized to take action."

With increasing exasperation, Mr. Ickes explained that wildfires were burning through grasslands and woodlands in the less-populated areas of Long Island. Many large estates (he emphasized "large" and "estates") were in grave danger. Senator Alphonse D'Amato of New York was on the scene and in direct contact with the White House. I assured Mr. Ickes that I would pursue the matter.

Associate Chief David Unger was acting chief in my absence from Washington. When I rousted him out of bed, he informed me that New York's state forester had requested a federal Incident Command Team to coordinate federal firefighting operations with townships and cities involved. Without an effective overarching command and control structure in place, there was a noticeable lack of fully functional coordination. In fact, the operation to this point had been likened to a Keystone Kops movie. As requested by the White House, a USFS Incident Command Team was on scene and/or en route.

In the meantime, Undersecretary James Lyons had gone to New York, at the instructions of the White House, to assuage the senator's concerns. D'Amato had been making appearances in firefighter regalia, giving some—especially the press—the impression that he was in charge and getting things done. To make matters worse, D'Amato had evidently called President Clinton or perhaps Vice President Gore to demand that C-130 air tankers be dispatched to drop retardant on the fires. Someone in the White House, without any consultation that we could discern, ordered the USFS to "make it so."

C-130s are very large airplanes originally built to transport troops and equipment for the military. Several of the giant planes, upon being declared surplus by the Air Force, had been converted to air tankers and turned over to civilian contractors to deliver fire retardant on forest and range fires. In addition, a few Air National Guard C-130s could, in emergencies, be converted to such use but, by regulation, could not be employed until all tankers operated by civilian contractors were already committed to combating wildfires.

I told the incident commander to be polite and respectful to the "politicals" but to ignore them completely in establishing control of the situation and fighting the fire. When I asked him if C-130s would be useful, he replied, "Absolutely not!" and spelled out the reasons. The closest Guard C-130 that could be outfitted as an air tanker was in South Carolina, and that would take at least a day to accomplish. Then it would take some time to load retardant (most commonly an admixture of ammonium nitrate and water). And it was unlikely that the standard retardant mix could be used on Long Island due to environmental concerns relative to municipal water supplies. That was being checked out. And even if the C-130 dropped only water from a few feet too low, it could knock a mansion wapper-jawed on its foundation. All in all, C-130s were the wrong planes for the job and, worse yet, would take the longest to get into action.

The commander respectfully reminded me that the Incident Command System was designed to negate just such political interference as was now emanating from Senator D'Amato and the White

House. A fully qualified incident commander was on the scene, had analyzed the situation, and was assuming command. When ready, he would request whatever air support he deemed necessary and suitable. And he did just that, making it clear that he did not need or want any C-130s.

Senator D'Amato, in the meantime, was giving the White House a continuing ration of crap, demanding to know why the C-130s "he had ordered up" had not appeared. The White House, in turn, was all over Secretary of Agriculture Espy, wanting to know when Senator D'Amato's C-130s would appear. Undersecretary Lyons got the next reaming and headed for New York to calm matters down after repeating the "request" for a C-130 to the USFS office in Washington.

To get the incident commander off the hook and cut him the slack to do his job, it seemed the better part of valor was to order up a single National Guard C-130 from South Carolina. The next morning that C-130 was en route to Long Island when the word came down that New York's Environmental Protection Agency had turned thumbs down on the standard retardant mix. So the C-130 turned back to South Carolina, offloaded the retardant, flushed the tanks, and filled the tanks with water. Now, as per regulations, pilots and crews had to take a break. Finally, early on the morning of day two, a National Guard C-130, loaded with water mixed with water, was on the way to New York.

By now, the incident commander had organized the firefighters and equipment from local, state, and federal agencies, and firefighting operations were under control and fully coordinated. D'Amato was livid. He had told the press, the television stations, and the people of New York that he had demanded that the White House provide C-130s—ASAP! Now here it was two days later, and no C-130s were flying low over Long Island with a plume of retardant stringing out behind to be filmed and broadcast on evening news and delivered courtesy of New York's very own Senator Alphonse D'Amato.

When the order was delivered to dispatch the C-130, a caveat was added: make sure the drop was from an altitude high enough to avoid any potential damage to structures or injury to firefighters. The public relations people told the politicians and the press when and where the drop would be made so the press could get pictures of Senator D'Amato and various local politicians looking up at the C-130 streaming water.

The C-130 dropped its load of water and turned toward its home base in South Carolina; after all, it might really be needed to fight wildfires somewhere else. In the meantime, the incident commander and his multiagency team—federal, state, city, and boroughs—had the wildfires on Long Island under control.

But the end was not yet. There was a demand from somewhere up the chain of command for an official inquiry as to why the USFS had not immediately complied with a direct order from the White House to dispatch C-130s to Long Island. The chief counsel to the secretary of agriculture was assigned to lead the investigation. Fortunately for all concerned, he was a calm, wise, and judicious man, complete with a sense of humor that proved handy in this situation.

Before the investigation began, I went to him with two messages. The first was that I was solely responsible for the failure to deliver C-130s as ordered (even though I was in Alaska at the time). The second was that the ordered investigation, if conducted honestly and thoroughly, would be highly likely to embarrass Senator D'Amato, the White House, and the secretary of agriculture.

I explained that the Incident Command System was designed to prevent inappropriate political intervention. Once an Incident Command Team took over, firefighters of all stripes were brought under a central command and the fires were brought under control as expeditiously as possible. The team had not requested or wanted C-130s. Further, dispatching a C-130 from South Carolina weakened national wildfire fighting capabilities while wildfire danger was high, and increasing, across the country. Investigators would find that the USFS had correctly followed the protocols. Further, it would be judged inappropriate for Senator D'Amato, the White House, and the secretary's office to have been involved. In any investigation it would emerge that I had given orders to stick with established processes and procedures—and took full responsibility.

Wisely, the investigation was abandoned before it began. The entire incident did not enhance my reputation with my bosses or the reputation of the USFS with some political operatives. But it was all carried out exactly "according to Hoyle," and the fires in question were quickly and efficiently extinguished. And perhaps most importantly, the senator, in his fire captain's regalia, got his picture in the papers. The Incident Command System, though challenged, remained intact and effective. All's well that ends well.

OPERATION LINEBACKER

In early 1995, a highly publicized, enhanced federal effort to impede illegal immigration from Mexico had focused, with significant success, on border areas immediately south of San Diego, California. As a result, the "coyotes" (smugglers of illegal aliens) had shifted operations eastward to where the Cleveland National Forest abutted the border.

Time-lapse aerial photographs revealed a spider web of rapidly developing trails that extended from the border with Mexico into the United States via the national forest. In several cases, it was reported that illegal immigrants who tried to make it on their own or had been abandoned by their coyotes might have perished from lack of water and food. Ranchers and other residents of rural areas along the border were vociferously complaining of thefts, trash, and exhausted, famished pilgrims showing up on their doorsteps begging for help.

I was informed that the USFS would, until further notice, provide fifty law enforcement officers to assist the U.S. Border Patrol working on the Cleveland National Forest. The effort was codenamed Operation Linebacker. I initially protested that USFS law enforcement officers were not trained as Border Patrol agents and that the USFS was already shorthanded in dealing with law enforcement on the national forests. My objections were "duly noted" and just as duly ignored. It was time to salute and say, "Aye, aye, sir." USFS officers were quickly assigned to Operation Linebacker.

The USFS's top cop, "Manny" Martinez, was waiting in my office when I returned. I relayed the orders. I allowed Manny opportunity and time to vent. Then I told him, as I had been told, that his objections and concerns were duly noted. Alone among all USFS personnel, command and direction of the agency's law enforcement officers were "stove-piped" so as to report directly to the chief—i.e., I was directly responsible for all law enforcement operations.

I instructed Martinez to establish liaison with the Border Patrol, prepare a memorandum of understanding between the agencies, arrange for the necessary training of USFS officers assigned to assist the Border Patrol, and establish protocols for equitable work details. Within ten days, USFS officers were on the ground, trained, appropriately equipped, and fully integrated into Operation Linebacker.

As the summer wore on, two types of complaints were being bucked up to my office regarding this arrangement: some USFS line officers objected to the agency's involvement in enforcing immigration laws, and a few Hispanic employees complained that the assigned officers were being "abusive" to Hispanics while assisting the Border Patrol. Martinez reported that, so far as he could ascertain, the complaints were unfounded.

I called the Border Patrol offices in Washington and California and was told that the USFS officers were doing an excellent job, especially in backcountry areas. However, that didn't put an end to the complaints. I invited several political appointees who were sniping about the "insensitive behavior" of USFS officers to accompany me on a review. They were, as is usual in "touchy" situations, "too busy." So to assure transparency, I invited two Hispanic female employees on the USFS's civil rights staff in the Washington office to accompany me.

Region 5 of the USFS (California) was a simmering hotbed of contention over the civil rights of women and minority employees. Involving USFS law enforcement officers with combating illegal immigration was, to say the very least, an internally sensitive matter. But the orders I had been given, wise or not, were both legitimate and nonnegotiable.

I met with upset Hispanic employees in the USFS's regional office in San Francisco and at the headquarters of the Cleveland National Forest in San Diego. Many employees seemed resentful that the USFS was involved in any way with enforcement of immigration laws. They were not mollified by my explanation that the president's office had ordered the agency to provide law enforcement support to Operation Linebacker.

We met with Border Patrol officials, who in general praised the performance and integrity of our officers involved in Operation Linebacker. I asked the Border Patrol commanders to come directly to me to report any incident of the USFS law enforcement officers' inappropriate behavior. We met with a number of local landowners and were shown examples of problems associated with the ongoing flood of illegal immigrants: trash, sanitation problems, cut fences, theft, and intimidation. Oddly enough, some were equally upset by the sudden influx of so many federal agents.

In a roadless area of the Cleveland National Forest, we joined a contingent of USFS officers assigned to backcountry trails being used by coyotes and their charges. A Border Patrol boat transported us across a lake to a trailhead where were met by a USFS officer who led us, on foot, several miles to a camp where we joined six additional USFS officers.

The facility consisted of nine nylon pop tents, a couple of nylon tarps strung up between trees to provide shade, a latrine, and a stash of water and fuel cans. We were assigned to tents and handed a paper sleeping bag (usually issued to wildland firefighters) and an inflatable sleeping pad. This was a "cold camp"—no outside fires and no lights were allowed lest they give away our position. Surplus military rations and hot drinks were prepared on small propane stoves.

One officer was always on watch and monitoring the radios. The others, having worked all night, slept in the shade of the tarps. It was hot—and a bit smelly. Water in five-gallon jerry cans had to be packed in and was too precious to be used for personal hygiene except for occasional sponge baths.

An hour before sunset, the officers were up, dressed in camouflage fatigues, and fed. Each carried an M-16 automatic rifle, a Beretta 9mm pistol, a small backpack, and a two-way radio. The team leader introduced his team and gave us a briefing. The Border Patrol operated primarily where roads and mechanized equipment let them cover more miles of border, handle more detainees more efficiently, enhance mobility and communications, and provide increased safety for their officers. The USFS officers, on the other hand, were quite comfortable operating in the roadless backcountry—on foot when necessary—and that is where they were assigned.

The USFS incident commander told us that coyotes evaded the Border Patrol by avoiding roads and leading their charges through the rugged desert mountains on foot. Coyotes routinely instructed their charges to scatter and hide when challenged by the Border Patrol. Then once the federal officers left the area, the coyotes gathered up those who had not been apprehended and continued on their way to the "promised land." Some coyotes, having been paid up front for their services, demonstrated no real concern for the welfare of their charges. Too often, the pilgrims in their keeping were simply abandoned—sometimes without food or water—to fend for themselves when apprehension seemed likely. It was a brutal, cruel, nasty, and sometimes fatal game best played by the desperate.

On a map, the incident commander showed us where his team had secreted electronic sensors that, when triggered, indicated the location of potential intruders and the direction of movement. Then likely intercept points were charted. I was impressed by the ingenuity, technology, and professionalism of the Border Patrol agents.

About midnight of our first night in camp, a sensor went off and then another and another. With minimal conversation, the USFS officers strapped on their gear, hoisted their packs, picked up their rifles, and took off up the trail. Several hours later, two officers with twelve illegal migrants in tow walked into camp. The prisoners were seated in a tight circle and told to remain quiet. Restraints were removed to allow them to use their hands and be more comfortable. They were judged too exhausted, cold, dispirited, and hungry—and too afraid—to run off into the darkness. The coyotes had told them that if they were apprehended, they would be severely beaten—or worse.

One obviously upset woman told the officers that the twin sister of her older child was still up on the mountain. The officers who were still looking for hiding pilgrims were informed via radio.

The two officers in camp asked if I would take over responsibility for the prisoners so they could join the search for the missing girl. This game of "adult hide and seek" had suddenly become serious— perhaps deadly serious. They gave the mother a half hour to rehydrate and rest. Then she went back up the hill with the officers to try to call the missing girl from hiding.

That left my two female companions and me in charge of the detainees. As he left camp, an officer handed me his semiautomatic Beretta pistol and two loaded clips and asked me if I knew how to use the weapon. I nodded.

We gave the prisoners our sleeping bags, water, and packets of rations, which they obviously needed far more than we. My two associates spoke fluent Spanish and interpreted my "TexMex" (a holdover from my ten years with the TGD) as I attempted to reassure our prisoners that no harm would come to them so long as they followed instructions and did not try to escape.

Finally, a young man asked me in heavily accented English, "You are the Forest Service, *sí*?"

I nodded, "How did you know?"

He had heard that USFS officers had begun to operate in the area. He continued. "The Border Patrol, sometimes, are not as kind as you have been." Was this fellow a coyote? For some reason, that seemed likely.

Shortly after daylight, a tired Border Patrol agent walked into camp and asked, "Chief, my boss wants to know if you would feel comfortable walking this group out to the lake. The patrol boats will meet you and take them into custody. We need all our officers to help look for the little girl that's still missing. He says if you don't want to take that responsibility, he understands. In that case, I will go with you to the boats."

I replied that we could handle the assignment. Besides, we were nearly out of water and rations. Our "guests" were footsore, worn out, cold, inadequately clad, in the middle of nowhere, confused, hungry and thirsty, and sick and tired of their whole adventure. I didn't foresee any trouble. The fellow I had tagged as a coyote was maybe five feet seven inches tall and weighed about 140 pounds. If he gave me trouble, I figured I would just sit on him.

We made the walk to the lake slowly but without real difficulty. It was a relief to see the boats waiting and turn over our charges to their keeping. When the boats pulled out, the young fellow who had spoken to me in English smiled, waved, and called out, "See you next week!" Ouch! I took it that he meant to poke fun at his adversaries in this very dangerous and expensive game. Were we really doing anything more than putting a minor, temporary dent in the influx of illegal immigrants?

Just then, one of the Border Patrol group told me that he had heard on the radio that the missing girl had been found. She was severely dehydrated and very frightened but otherwise in good shape. All of those big, tough officers and I were greatly relieved.

When I got back to Washington, I told the secretary of agriculture that, so far as my team and I could determine, the USFS officers assigned to Operation Linebacker were doing a superb job under difficult circumstances. Those officers, in my opinion, richly deserved commendation, which I ultimately made happen with the secretary's support. I heard no further complaints. The USFS officers had faced up to a challenge and enhanced their reputations in the process. And most certainly, I increased my understanding of and standing with them.

ROADLESS AREAS: PERPETUAL BONES OF CONTENTION

By 1993, it was increasingly obvious that the USFS had simply hit a wall in striving to meet the timber target of nearly 13 billion board-feet per year mandated by a long succession of Congresses and administrations. Such troubles started with the National Forest Management Act of 1976, which mandated national forest planning. Planning operations had been completed under intense scrutiny and never-ceasing political pressure from President Ronald Reagan's appointee, Undersecretary of Agriculture John Crowell, who was formerly a lawyer for the timber industry. I came to think of Undersecretary Crowell as the personification of the fox in the henhouse as he pushed incessantly to increase annual timber harvests to levels that many, if not most, USFS professional thought to be legally unreasonable as well as unsustainable.

In the opinion of some analysts, millions of acres of the national forests had been inappropriately classified as suitable for timber harvest in what some called "a politically dictated fit of irrational exuberance." Many forest supervisors were shying away from building roads into areas where they believed that timber cutting made no economic, ecological, or political sense.

Still, those same supervisors, with rare exceptions, did not bite the bullet and make serious attempts to adjust their individual plans to remove such areas from the "suitable for timber harvest" category. Dealing with that ever hotter political potato was, year after year and decade after decade, put off for successors to handle.

The areas absent engineered roads came to be referred to—for purposes of the coming internecine conflict over federal land management—as "roadless areas." Environmentalists, fish and wildlife biologists in state and federal agencies, and recreationists (including hunters and fishers) increasingly spoke out in favor of retaining such areas in roadless condition. In 2011, I published a two-part piece in *Fair Chase* magazine on this same subject titled "The Future of the National Forests: Who Will Answer an Uncertain Trumpet?"

When I became the USFS chief, I tried to bring questions over the management of roadless areas to a head. I gave regional foresters and the forest supervisors under their supervision two options. They could commence, forthwith, building roads into presently roadless areas classified in forest plans as "suitable for timber harvest" and strive their best to meet the timber harvest targets. Or they could, with dispatch, set out to amend individual national forest plans to reclassify such areas as "roadless"—and reduce anticipated timber yields accordingly. I was naïve enough at that point to believe that if I gave a direct order, it would be automatically and vigorously obeyed. I was wrong.

Most forest supervisors did neither. Caught between a rock (meeting the planned timber targets) and a hard place (building roads that made no economic or ecological sense under public scrutiny), they continued on their current paths. That, in essence, left increasingly more controversial decisions to their successors. Even so, timber harvest rates continued to decline rapidly. The status of roadless areas was a festering boil on the collective ass of the agency. Sooner or later, one way or the other, that festering boil would have to be lanced.

By the 1990s, current national forest managers at every level had been left holding the proverbial bag after decades of harvesting the "good stuff" and the "easier stuff" while leaving the "lesser stuff" and the "tough stuff" for later. More and more forest supervisors and district rangers understood that they were increasingly dealing with "tough stuff" and would be dealing with "ever tougher stuff" as time passed. That recognition put roadless areas front and center in the growing public debate.

In 1994, Congress addressed the question of roadless areas by simply refusing to fund any new road construction. It took considerable political effort to salvage even a much-reduced allowance for road building and maintenance. As chief, I was content in the short run to avoid building roads into roadless areas—and to delay the determination of the fate of individual roadless areas in the next round of forest planning. The Clinton administration exhibited no will to lead on the issue. From the political view, there was no way to finesse the issue—it was damned if you do and damned if you don't. Environmental warriors seized upon the roadless-areas issue, and the conflict quickly intensified. At that time there were some 380,000 miles of "mapped roads" in the 153 national forests and an estimated 60,000 miles of "unmapped" roads—a total of 440,000 miles of roads.

The upcoming round of forest planning was predicted to be protracted, contentious, expensive, and unpredictable. I expected, though, that individual decisions would be made for each national forest. It was further anticipated that economic and ecological assessments, coupled with building public opinion in support of roadless areas, would result in the vast majority of roadless areas being retained. In my opinion, the USFS had already built roads into the vast majority of areas where roads made any economic sense.

Furthermore, roads require routine maintenance. The Knutson-Vandenberg Act of 1930 authorized the USFS to retain 10 percent of timber sale receipts for management purposes—including that of roads. So the dramatic reduction in timber sales, beginning in the 1980s, reduced the resources needed for adequate road maintenance. It was a classic "do loop"—more timber sales meant more roads while providing funds needed to maintain the road system as it grew inexorably. Some called it a new variety of Ponzi scheme that would lose viability if and when timber harvest rates dropped significantly, which seemed inevitable—and likely sooner rather than later.

When my tenure as chief ended in 1996, my successor was Dr. Michael "Mike" Dombeck, an old-hand USFS fisheries biologist who had been "loaned" to the Department of the Interior as an assistant to the director of the BLM, where he had spent three years as acting director of that agency. Dombeck was highly regarded, with good reason, and to his great credit, he took on the roadless-areas issue from the top down—in one very big bite—outside the traditional land use planning and land allocations processes. He simply declared, through appropriate processes, that identified roadless areas would remain in that status.

President Clinton, through USFS rule-making processes, retained all the roadless areas greater than 5,000 acres in area in roadless status. The decision was to be part of his "environmental legacy." The environmental community as a whole and many in the hunting and fishing communities were ecstatic. It seemed, so far as the national forests were concerned, to signal an end to this contentious issue. That decision was expected to yield significant savings in time, money, and political capital. Chief Dombeck had the will and political skills, at least for the time being, to put the issue of roadless areas to bed. But for how long?

Then a significant number of those involved in timber extraction, mining, motorized recreation, and livestock grazing rose up in opposition. Many believed the effect would be the creation of *de facto* small or "pocket" Wilderness Areas. Some interpreted this as an end run around established laws, rules, regulations, and gentlemen's agreements. Legally, only Congress can designate Wilderness Areas. However, roadless areas are not Wilderness Areas by any stretch of the imagination and are not managed under the same rules. Roads can make all the difference in the management of a block of land.

Four years later, this maneuver would become an issue in the presidential election of 2000. Vice President Al Gore, who held "the environmental portfolio" in the Clinton administration and supported Chief Dombeck's decision, was defeated in his bid for the presidency. Dombeck retired and took an academic appointment at the

University of Wisconsin. There was a new gang in town under the directorship of President George W. Bush.

In 2004, a federal judge in Wyoming ruled that the process used by the Clinton administration to protect roadless areas was illegal. By failing to appeal, the administration accepted that decision. So a new process was put in place to determine the fate of roadless areas. The Bush administration, likely coached by the new quite skookum Undersecretary of Agriculture Mark Rey, announced that, unless state governors chose to opt out, the issue would be reconsidered on a state-by-state basis.

States were allocated federal dollars to cover the costs of arriving at their decisions. In theory, the USFS would have the final say. It was a new, quite clever political ploy that should have appealed to the president's constituency relative to timber harvest levels from the public lands. Tossing this hot potato into the hands of governors made them feel, in the short term, trusted and appreciated. However, as some governors quickly understood, they would have to take some of the flak from those who opposed the ongoing losses of increasingly rare roadless areas.

Whether one agreed with, appreciated, or even understood the actions of the Bush administration concerning roadless areas, it was fork in the trail for management of the national forests. Governors seemed to have new powers to influence such management with no associated liabilities to fund road construction, maintenance, or management. And, just as clearly, attempts to remove such areas from protected status were apt to set off a political donnybrook.

Perhaps governors should have very carefully looked a gift horse (or was it a Trojan horse?) in the mouth. Some governors rather quickly figured out the game—and the potential political consequences. The ball was, to some degree and as was promised, in their courts.

Why should hunters and fishers care? Because roadless areas generally represent some of the best remaining, relatively pristine watersheds and fish and wildlife habitat on public lands. And the bad news is that construction and maintenance of roads commonly does absolutely nothing to improve that fish and wildlife habitat. In fact, roads almost inevitably have a deleterious and sometimes dramatically negative impact. But the good news is that roads improve access for hunters and fishers and recreationists.

However, motorized recreationists—with four-wheel-drive vehicles, trail bikes, and the like—present increased management problems because they often degrade watersheds and create the need for enhanced road management. And such users, in general, pay no fees to federal land management agencies for increased management to compensate for the consequences of their activities.

Where would the funding come from to institute and maintain road management? Funds were already grossly inadequate to maintain and police even existing roads, and it was foreseeable that the situation would worsen. An old saying—"When you find yourself up to your neck in crap, quit digging!"—seems good wisdom in this case.

There are good reasons for roadless areas. First, the USFS had already built roads into nearly every nook and cranny of the national forests where they made any sense—and in many areas where, in retrospect, they made little or no sense. Second, road standards and costs have steadily increased for good environmental reasons, and that seems likely to continue. Honest and thorough cost/benefit analyses relative to more road construction make it highly unlikely that building and maintaining more roads would be judged both economically feasible and ecologically rational. Third, administrations and Congresses have been consistently reluctant to provide for adequate upkeep of even the current road systems. Fourth, money for road maintenance is not apt to increase since timber receipts, which have largely paid for road construction and upkeep, have continued to dwindle. Fifth, burgeoning federal deficits make it unlikely that adequate funds for maintenance and management of roads will be forthcoming from general appropriations. And, sixth, building new roads into roadless areas is not a fight upon which astute political appointees and elected officials will likely want to expend precious political chips. More roads will, in general, only come with considerable political pain.

As my old hunting *compadre* Bill Brown said to me during one of our many wilderness excursions, "Remember, managers don't make *roadless* areas; they make *roaded* areas. Building a road is a permanent decision; there is no going back. It will be an ever more rare and precious experience to go where there are no roads and to experience the 'great silence' without the sound of internal combustion engines."

Those who demand roads into the last roadless enclaves, when the vast majority of federal lands are already crisscrossed with enough roads to reach to the moon and back, need to ask, "How many roads are enough?" Would a single governor recommend building roads into the remaining roadless areas in the national forests in their state—and maintaining them in perpetuity—if the necessary funds for construction and maintenance and management came from the state's treasury? Really?

I once expressed such feelings in a letter to a newspaper editor. A critic eloquently responded that, now that he was old and increasingly infirm, it was his "right" to be able to visit the haunts of his youth in roadless areas via motorized transportation. I, and others like me, were bent on denying him his "rights." I responded that I was near eighty and so crippled up from old injuries that I could no longer ride a horse or walk more than a few miles. I miss my old haunts and the memories and ghosts that reside there. But I still didn't want roads built into those precious places. Those places, and my experiences there, are fresh in my memory and will reside there so long as I live and have good sense.

I want and cherish the same for others—young, strong, and vigorous others—to have the same opportunities, maybe even the "rights" I exercised in my younger years to visit the few remaining places on the public's lands where roads don't go—not yet. I want my grandchildren and their children's children to have the chance to see, visit, feel, and know such places. And, most of all, I want them to be able to know, hear, and feel the great silence.

At some point the discussion should be about "them"—those yet to come generation after generation. Do they not have "rights" to know the great silence earned by simply getting there on their own power? It is way past time to let this largely symbolic issue go and move on to better expenditures of time, money, and political energy.

As a nation, we have come to a major fork in the trail relative to the future of roadless areas in our national forests. We seem to be shuffling from one foot to another as we contemplate those forks in the trail. I remember the admonition tacked onto the sign at the fork in the road where the owl sat in the picture left in my office so long ago. Once we build a road, there is no looking back, and there is "No longer an option." So now, while we have both time and opportunity, we desperately need to look back and ponder what we have done—and contemplate what comes next.

Looking back, what are the actions of which we are proud? What do we regret? Then we can look forward and contemplate the next fork in the trail, better informed as to our personal and societal decisions. After all, if we build no new roads into roadless areas today, the decision can be revisited—over and over, generation after generation—as times and circumstances change. It is well to remember, and ponder, the words on that post at the fork in the trail: "No longer an option."

CHAPTER 6

A TIME TO TEACH AND TO REFLECT:
MONTANA, 1997–2005

On the last day of December 1996, after thirty years of federal service, I retired as the USFS's senior scientist and chief of the agency. On the first day of January 1997, I became the second Boone and Crockett Professor of Wildlife Conservation at the University of Montana in Missoula, a truly major fork in my trail. And in a miracle of circumstance, red-haired Kathleen Connelly had come into my life when she was promoted from a job in the Department of Agriculture to be the deputy chief of Administration for the USFS. We became friends—then more than friends. We were married after we both retired from government service, and once again I was blessed with a bride who is that rare combination of a loving companion, sweetheart, and best friend—and a bright-as-a-button, accomplished professional to boot.

THE JUXTAPOSITION OF HEAVEN AND HELL
One of my duties as the Boone and Crockett Club's endowed professor was to oversee the research carried out on the Theodore Roosevelt Memorial Ranch (TRM Ranch) on the eastern front of the Rocky Mountains.

When William "Bill" Spencer was president of the Boone and Crockett Club (B&C), he led the charge to bring one of the oldest conservation organizations in North America, founded by Theodore Roosevelt and ten associates in 1887, back into a position of major influence in conservation of natural resources. His dream included the B&C's owning and managing a critical piece of real estate on the Rocky Mountain Front in Montana. He visualized, simultaneously, preserving the property from subdivision and development; demonstrating that "practical" ranching is compatible with protecting wildlife and its habitats; providing a locale and facilities for conservation education of both children and adults; and establishing a "laboratory"—and a retreat—for graduate students in training to be tomorrow's leaders in conservation.

Spencer found a suitable location in what he called "the Serengeti of North America," which B&C purchased in 1986. This almost-magical place lies near the tiny town of Dupuyer, Montana, where the Great Plains and the Rocky Mountains seem to collide. The ranch adjoins the USFS's Bob Marshall Wilderness and the State of Montana's Blackleaf Wildlife Management Area. Taken together, the three properties provide a complete year-round habitat for mule deer, elk, grizzlies, black bears, cougars, and other charismatic species, more recently including wolves.

There animals and plants of two distinct ecological types, the Rocky Mountains and the Great Plains, intermix and interact in fascinating ways. The lower elevations of the mountains are dominated by conifer forests, the plains by grassland and shrub habitats. The foothills between, a transition zone, are a mixture of both types of vegetation. Conifers dominate the north-facing slopes, hardwoods along the stream bottoms, and open grasslands on the ridges and south-facing slopes. The scenery is as breathtaking and varied as the ecology is fascinating. The Rocky Mountains loom above the ranch. Especially at dawn and dusk, it is clear why authors have called Montana Big Sky Country and "the Last Best Place."

Like many USFS personnel who were Professional Members of B&C, I had never visited the TRM Ranch. I arranged a "meet and greet" with the ranch manager just before spring semester at the University of Montana began in the third week of January.

Driving to the ranch from Missoula in the midst of winter can be challenging, sometimes thrilling, yet always beautiful. The po-

A view of the East Front of the Rockies from the Theodore Roosevelt Memorial Ranch, Dupuyer, Montana.

tential difficulties in making the drive include mountainous terrain, sometimes very windy conditions, snow and ice, and the fact that many of the roads—in true Montana style—have very skinny shoulders or none at all. By Montana custom, small white Christian crosses—one cross per fatality—mark the roadsides where fatal accidents have occurred. Those crosses are meant to simultaneously commemorate the dead and caution the living.

Montana had absolutely no speed limits on rural roads when I first moved to the state at the end of 1996. Speeds were to be "reasonable and prudent." That, like beauty, lay in the eye of the beholder. And to top off this devil's brew, it was legal to drink and drive so long as the driver could pass a sobriety test. Roadside bars had drive-up windows where drivers with the hankering for an alcoholic drink could get one for the road. Fortunately, that is no longer true.

But in compensation for the hazards, the drive from Missoula to the TRM Ranch was surely one of the most beautiful I had seen in all of North America. My route took me east from Missoula along Interstate 90 through a gap in the mountains where the Blackfoot River flows on its journey to the Pacific. After a few miles, the turn onto Highway 200 leads through mountain-flanked valleys toward the spine of the Rockies. In some ninety miles, the climb out of the river bottom begins, and ten miles later one crosses the Continental Divide at Roger's Pass—5,610 feet above sea level. Now trees become ever more sparse as the road descends onto the Great Plains. There the only trees are scattered along the intermittent stream courses. From frequent vantage points, it seems that one can see forever.

If you squint, it is easy to imagine that the specks of cattle in the distance are buffalo. A little over a century ago, hunting parties of Native Americans from tribes that lived west of the Continental Divide annually followed this same trail into buffalo country to harvest meat and hides. Just after crossing the Dearborn River, which eventually joins other rivers on the way to the Gulf of Mexico, one turns north on Highway 287, which leads to Augusta. Off to the west, the mighty shoulders of the Rocky Mountains loom high above the rolling grasslands.

The Old North Trail, which bisects the TRM Ranch, follows the eastern front of the Rockies north-south from Alaska to Mexico. Some believe the first humans who crossed over from Asia to North America during the last Ice Age established this trail in colonizing North America. When one stands on that trail, still marked here and there with what are said to be travois tracks, it is not difficult to summon up visions of the various travelers at various times over the last thousands of years. Such thoughts combine with the "Rocky Mountain high" to shrivel egos—at least mine—to a more appropriate size.

Elk and mule deer are commonly seen on the southern and western exposures of the open grasslands. Antelope are occasionally seen. And one knows that grizzlies and black bears are out there somewhere, in hibernation, waiting for the warmth of spring to coax them out of their dens. Cold winds blow dry snow across the road in swirls that can remind one of sands blowing in the desert winds. All sights and sounds can, to those in the right frame of mind, speak of beauty, peace, and tranquility while simultaneously reminding of wildness and providing a whiff of danger. It is the very subtle complexity of the admixture that sweetens the experience.

Then I noted—a quarter mile off the road—an enclosure of an acre or so made up of a eight-foot-high wire-mesh fence topped with barbed wire. Though there was no sign on the fence, I knew that it enclosed the entrance to a silo that housed an intercontinental ballistic missile tipped with multiple nuclear warheads. The missile was programmed to deliver those warheads to preselected targets in nations *de jure* considered a threat to our national security. Huge flat concrete doors, which would slide open just before launch, precluded disturbance of the sleeping beast. As I continued on my way, I saw more missile silos placed unobtrusively in swales in the landscape.

The silos were located in the shadow of the Rocky Mountains because—in the days before satellite technology—if the missiles sleeping in the silos were launched, they would be well on their way before they could be detected, due to the shielding effect of the mountains coupled with the curvature of the earth. And in the event of a nuclear accident or a strike launched by an enemy to destroy or disable the missiles at rest in their silos, human casualties would be "minimal" due to the isolation.

I counted the silos—the ones I could see—knowing that there were many others that I could not see. I tried to compute how many nuclear warheads would be on their way if all the missiles, many with multiple warheads, were simultaneously launched. It made my head swim, my gut churn, and my heart ache. Then I pondered all the additional long-range missiles, short-range missiles, nuclear artillery, long-range bombers, medium-range bombers, and fighter bombers poised around the world equipped with nuclear weapons. Then I re-membered to add the warheads aboard aircraft carriers, missile-car-rying cruisers, and nuclear submarines. I knew of all of those things and, as an Air Force reservist, had for a very long time.

But encountering missile silo after missile silo in this most beautiful and bucolic of settings grated on my mind—and on my very soul—much like fingernails on a chalkboard. Perhaps this was the ultimate expression of the blackness and madness that sometimes seems to lie buried deep in the collective heart of humankind. Un-fortunately, that madness festers and boils to the surface from time to time. Clearly, if all these missiles were ever launched—matched by those of the "other side"— they would indeed announce the arrival of the long-foretold Armageddon.

To add to my nightmare, I knew that somewhere out there across the Pacific—in China, the Soviet Union, and North Korea—similar silos held similar missiles that could be loosed in first attack or in response to an attack. The sheer madness of my visions was made even more bizarre by the awesome austere, serene beauty of the coun-tryside through which I passed.

I thought, naïvely, that if all the leaders who held sway over such weapons could see the potential horror while simultaneously feeling the sublime peace that this landscape exuded, they could not help but recognize the obscenity of the death and destruction lying, at least for now, sleeping in the silos. There the specter waits, patiently and carefully tended—but for how long? Only a series of electrical impulses is required for them to spring to life and send them on their way to deliver death and destruction.

As I departed the tiny town of Choteau and headed up the road to the TRM Ranch, the infamous winds that from time to time blow down off the Rocky Mountain Front suddenly struck, bearing heavy snow. It took me well over an hour and a half traveling in the blinding snow to cover the last thirty miles.

When I arrived at the ranch's guesthouse, the driveway and sidewalk were covered with ice and blowing snow. I was inappropriately shod in cowboy boots with leather soles. Upon stepping out of the car, I was, to my surprise and consternation, blown off my feet and slammed face down in the snow. Using the car as a crutch, I got up and immediately went down again, this time on my fanny. I crawled around to the trunk, opened it, and retrieved my duffel bag. I shut the trunk and crawled to the front door, dragging the bag. Fighting against the unrelenting wind, I managed to stand up and pry open the storm door.

When I turned the knob on the front door, the wind tore it out of my hands and slammed it open. I threw my bag inside and followed on my hands and knees. I had been out in the gale-force wind and snow for less than fifteen minutes and was as deeply chilled as I had ever been in my life. I thought that, if the temperature was ten degrees below zero and the wind was gusting to sixty miles per hour, the wind chill must have been minus . . . well, a whole bunch, maybe two or three whole bunches.

When I went to bed, I tossed and turned and could not sleep. I had never heard such a wind, howling steadily and sometimes literally screaming. The brick house even seemed to shake. It was a prescription for twilight sleep and nightmares. The surreal contrast of the beauty of the countryside with the potential horror in those missile silos kept coming back into my mind. It was hard to distinguish thoughts from dreams—and dreams from fears.

Just at daylight, the wind abruptly died down; it was deathly still and so very quiet. The sudden peace was rendered stunningly beautiful by a flaming sunrise in a clear sky. Then the sun peeked up over the Great Plains and lit up the Rocky Mountain Front, starting with the bare granite peaks and working down. Mule deer and Rocky Mountain elk appeared, seemingly from nowhere, on the east-facing slopes to paw through the newly deposited powdered snow in search of their "daily bread," as their ancestors had done for untold centuries. They were living on slim pickings now—and on their body fat stored in the more bountiful days of late summer and fall. Many of the adult females were pregnant. For them, the age-old annual race was on between starvation and the arrival of another spring—and another generation of their kind. Some would die in the meantime. Yet life would go on. I pondered that the deer and elk didn't know about the nuclear warheads quietly residing in this juxtaposition of extant heaven and potential hell. I concluded that they were all the better off for their ignorance.

In May of 2010, the United States and Russia announced an agreement to reduce their stockpiles of nuclear warheads—over the next seven years—to not more than 1,550 each, down from the previous maximum of 2,200. Just where those reductions would occur was not specified. At that moment, 450 Minuteman III missiles, each with several nuclear warheads, were scattered and asleep in their silos across the countryside of eastern Montana, North Dakota, and Wyoming.

Some business leaders and politicians in all those states immediately began lobbying their congressional delegations to keep their missile silos loaded, armed, and manned. Why? Negative impacts on employment and local economies that would come along with any reductions. That was the bad news to accompany the good news. Those missiles maintained in their silos were, are, and likely will remain someone's bread and butter. It is indeed, I suppose, a foul wind that blows no good. Here lies perhaps the most significant fork in the trail ever faced by mankind. Having come this close to Armageddon, can mankind ever retreat? It was a crystal clear, still, very cold morning on the Rocky Mountain Front—a new day filled with possibilities. And we have one more day to face up to mutual problems and seek solutions. After all, that is what forks in the trail are all about.

THE GRIZZLY AND THE JOURNALISTS

Among the facilities at the TRM Ranch is the Elmer E. Rasmuson Wildlife Conservation Education Center. B&C encourages other conservation organizations to use the center for meetings or training sessions. In the summer of 1999, the facility was rented to a group offering training on natural resources issues in the western United States to practicing journalists from across North America. The leaders for the session were Frank Allen, former dean of the School of Journalism at the University of Montana, and his wife Maggie.

I had been invited to spend a half day with the group, who had been at the center for a couple of days when I arrived. When I drove in the gate to join them, I recognized two graduate students from the university walking down the lane. They were studying grizzly bears. At that time, the state of the art for such studies was placing radio collars around the necks of bears; the collars gave out an individualized signal, and biologists used wands to detect the signal through earphones—and get a bearing. Once two operators had a bearing, triangulation revealed the bear's location. Now archaic, it was golly-gee whiz-bang technology at the time. I rolled down my pickup's window. "Hi, guys, what bear are you receiving?"

"It's a sow that's been collared for a couple of years—old No. 104. She has a cub with her. She's right down there a half mile or so along the creek. She's a 'good ole bear' and never causes any trouble or gets anywhere close to people if she can help it."

We passed the time of day before I headed on down the ranch road to the education center. On the way I encountered two journalists with fishing gear in hand. They were on their way down to Dupuyer Creek. I introduced myself. "Fellows," I said in an exaggerated Texican drawl, "y'all need to be real careful now; talk loud and make some racket. I suspect there's a sow grizzly down along the creek, likely with a cub in tow."

They looked at each other quizzically and then back at me. I could tell what they were thinking, "Yeah, sure. And just how did you figure that out, old-timer?" They no doubt suspected, with some justification, that I considered them greenhorns and was jerking their chain.

The devil made me do it. "Well, I've been a wildlife biologist for nearly a half century. Over that much time, really good biologists just sort of develop a sense about critters they study. You add up what you know about how bears act and the type of habitat they favor at different times of year and different times of day under various weather conditions. And, out of all of that, you get feelings that are more often right than wrong. You boys heed my words, now. Y'all be real careful."

They rolled their eyes and gave each other a look that challenged the veracity of the speaker, one who is clearly an old fart, into question. I chuckled to myself, thinking that they, being newspaper men and all, had probably developed pretty good bullshit filters. They resumed their journey. I resumed mine.

The rest of the group were sitting on lawn chairs at the guesthouse sipping drinks to celebrate the end of a long day of lectures by an array of speakers—environmentalists, conservationists, loggers, ranchers, federal and state wildlife experts, and journalists. I introduced myself and allowed as how I had met a couple of their comrades on their way to the creek and warned them about a sow grizzly and cub. Asked how I knew about the bears, I stayed with my line of malarkey.

I was still on my first cup of coffee when the two journalists came huffing and puffing up the hill. They were sweating, short of breath, and obviously excited. They looked at me. "We *did* see a grizzly—and a cub! Just where you said!"

The journalists were having a great time with their grizzly encounter. My prestige—and the respect accorded me—grew with the ongoing consumption of tall cold ones.

The next morning, I gave my two-hour lecture with no further bullshit about grizzly bears, then I loaded up my gear and started home. At the gate, it came to me that I had never told my audience how I knew about the grizzlies. I could imagine one or more of them writing up the story exactly as I had told it. I visualized a headline: "Former USFS Chief's Instincts Prevent Potentially Fatal Bear Encounter"—or some such misinterpretation of reality. I turned around and went back to the center to correct the record. When I did, some laughed, some smiled, and at least a few seemed embarrassed. But, all in all, my confession was welcome. A couple of the journalists, maybe more, had already written up their notes for future use.

I was happy. One Grizzly Adams was enough. A Grizzly Jack—

especially one who was a total sham—would have been one Grizzly Somebody too many.

GAME FARM "TROPHIES"

In the fall of 2003, I was seated in first class on a commercial flight (a privilege sometimes accorded to those who fly far too much) from New York to Missoula. I was reading an issue of the *Journal of Wildlife Management*, published by The Wildlife Society. My seatmate assumed, reasonably enough, that I was a wildlife biologist and introduced himself. He had recently been given a copy of *North American Elk: Ecology and Management*, published by the Wildlife Management Institute, and he recognized my name as an editor-author. That turned the conversation to elk—and the hunting thereof.

Before long, he retrieved his briefcase and fished out a wad of photographs. Most showed him in various poses with a bull elk he had recently killed. The bull's massive antlers sported seven points to a side. His "professional outfitter-guide" had scored the antlers using B&C's well-accepted technique. The score verified that this was, indeed, one of the largest set of elk antlers I had ever seen.

His "hunt" had taken place under perfectly legal circumstances on an "elk ranch" enclosed by a high fence. The animal had been born and raised on that licensed elk farm until it was five years of age, when it was sold as a breeder bull to the owner-operator of the "hunting preserve" where he had "hunted."

Initially, the bull had been confined in the breeding pens and fed a high-protein diet laced with nutritional supplements such as protein, calcium, and phosphorus. The intent was to produce inordinately large antlers. As the only male in the breeder pens, he serviced the cow elk turned into the pen with him while being spared injuries that could have resulted from clashes with other bulls.

At age 8½, the bull's owner determined that the animal had served his purpose as a breeder. And, perhaps just as important, the bull was likely at the peak of his antler development—and his monetary value as a "trophy." My seatmate verified, with an insider's wink, that he had paid "well into five figures" for his "trophy." He had seen pictures of the big bull, along with an estimated score on the set of antlers, before he booked the hunt. Once the deal was struck and the "sportsman-buyer" was on the way, the bull was transferred to a "shooter pen"—an enclosure large enough and with enough vegetation to provide an illusion of natural circumstances. Yet it was small enough to preclude any chance of the client wounding and losing the animal.

He had several choices of hunting experiences. He could shoot from a raised blind, a small enclosed "room" built on stilts some twenty feet off the ground and overlooking the feeder that provided the big bull his daily bread. The bull would be deprived of supplemental feed from noon the day before the so-called hunt, so he would come quickly when he heard the automated feeder go off and scatter pelleted feed. Or the client could shoot from the "safari wagon," a pickup truck with swivel seats and padded "shooting rails" over which he could rest his rifle when making the killing shot. Or, if he chose, the client could "stalk" the bull on foot, backed up by a guide who would finish the job if that proved advisable or necessary. A close-up shot was assured, as the pen was relatively small, the bull had minimal fear of humans, and the "guide" was there to approve the shot—and assure the kill if that became necessary.

My seatmate, true hunter that he was, disdained the blind and the safari wagon options. He chose to hunt on foot. He boasted that he had "finished the job in less than a half hour" with "one well-placed round" from his .300 Weatherby rifle with a 6-power scope.

He showed me photos of the trophy room he had added onto his house a decade earlier. The room had a high ceiling and teakwood-paneled walls. The mounted heads of his trophies, taken over the years in various places around the world, hung in several rows. Individual spotlights mounted on the vaulted ceiling illuminated each of the mounted heads: several white-tailed deer, mule deer, and elk. In addition, specimens of exotic species of deer, including axis, sika, red, and fallow deer, all with massive antlers, also adorned the walls. From his descriptions, I concluded that most if not all of those "trophies" had been killed on so-called "canned hunts," similar to the one he had just described. He did not say specifically what he had paid to kill the animals, but he allowed that the fees ranged upward from $2,500—some quite significantly upward.

The hour was getting late and our conversation, one-sided as it was, was winding down. The cabin gradually darkened as one passenger after another turned off their reading lights and drifted into what passes for sleep in an airline seat. I put my seat back, closed my eyes, and pondered what I had just heard and seen.

It was obvious why entrepreneurs would enter such a business: it was an economic niche that could be exploited for significant profit. It was a relatively new way to increase revenues from the ownership and management of land and creatures on that land—a lesson straight out of Business Management 101 taught in any college. Such operations could take place on acreages of insufficient size to otherwise turn much of a profit.

Clearly, my seatmate and others like him could be well satisfied with the arrangement. After all, he had kept coming back, checkbook in hand, and bringing along hunting companions. His "hunting" activities were in full compliance with all applicable laws. The landowner made money, the folks he employed made money, and the taxidermist made money. And the "hunter" was, most certainly, happy and looked forward to his next "hunt"—a cross between free enterprise and capitalism and "hunting" at its best! After all, money does talk.

So why did I find the show-and-tell, while educational, so distasteful?

To my mind, the pictures were not images of trophies at all but rather phony trophies—carefully manufactured facsimiles of what hunting trophies should be. They were purchased at what might be called a "trophy store" without even a reasonable facsimile of what I thought of as fair-chase hunting thrown in. The mounted heads on the wall of the buyer's trophy room were but decorations symbolizing nothing beyond skill in intense animal husbandry on the part of the seller and an ability and willingness to pay—and pull a trigger—on the part of the buyer.

Other such operations that I have known over the years in various places around the world were closer to maintaining an image of what might be called the "natural world." But all were tightly focused on the production of an intensively managed crop of male animals with unnaturally large antlers or horns of recognized game species to satisfy clients who paid top dollar to kill these monsters and pass off the freakishly large antlers as trophies. Maybe, I thought, such animals should be referred to as "Frankendeer" or "Frankenbulls," with my sincere apologies to Dr. Frankenstein.

I thought of the three primary factors that influence antler/horn growth: age, nutrition, and genetics. These variables are manipulated by managers of wild herds of the deer family and trophy farmers alike. Yet such circumstances as my seatmate described provide no genuine "trophies" and only a vague facsimile of "hunting." To my mind, they do not even measure up to the pejorative term "canned hunts." As no real hunt is involved, such are better described as "canned shoots." Using words like "wildlife," "hunter," "game," and "trophy" to describe these situations border on travesty.

The slaughter could just as easily have taken place in an abattoir. No real skill in stalking was required, no honored traditions were upheld, no honor or respect was accorded the prey, and but little skill was involved in dealing death. Even the very flesh of the animal was not cared for by the hunter and may not even have gone to the sustenance of the hunter's family. Meat from such animals could have just as well have come from the butcher's shop.

These were not new thoughts. Aldo Leopold observed that the values of wildlife and of hunting were inversely related to the artificiality of the surrounding circumstances. Could Leopold even have conceived of the advanced level of "artificiality of circumstance" described by my seatmate? Unless real hunters address, abhor, and condemn such extreme practices, the tide of public opinion may well turn from acceptance of hunting—however grudging—to dedicated opposition. And maybe it should. This developing fork in the trail is one that the hunting community would do well to avoid.

I had fretted enough. It was time to sleep.

JUST WHAT IS A TROPHY?

B&C is best known among North American hunters for devising techniques to measure the "trophy value" of antlers and horns of big game animals. It maintains periodically updated records books, wherein the most magnificent of those trophies are authenticated, re-corded, and recognized through a standardized scoring system.

When I was identified to strangers as the B&C's endowed professor at the University of Montana, ensuing conversations frequently led to a discussion of trophies. Because of my profession as a wildlife biologist and the connections it provided, I enjoyed, over six and a half decades, many opportunities to hunt and kill what would be considered trophy animals. Yet I have killed only a few animals that would make the records books. When I did, I gave them to various institutions for display. But now that I was the B&C's endowed professor, I was more and more forced to come to grips with the subject of trophies—what they should and should not be. Webster's dictionary seemed a good place to begin. "Trophy" is defined as "something gained or given in victory or conquest especially when preserved as a memorial."

I was disgusted to think that definition would be used to describe a hunting trophy. The hunter wins no "victory" over an animal killed. How can killing an animal produce a "victory," when the animal cannot even conceive that there is a competition under way—with rules, no less? The only victory for the hunted is to escape death, live another day, and reproduce. Further, no superiority over other hunters is necessarily demonstrated in taking a trophy-class animal. After all, luck and equipment can be and often are as important as hunting prowess.

Each hunter should contemplate, over and over again as years pass, just why they hunt. The introspective hunter must deal with the inevitable intertwining of motivation, knowledge, ethics, skill, feelings, traditions, and instinct. For example, in my musings, I have come to wonder whether hunter and prey are really independent entities, or if they evolved together over millennia and so cannot be separated? Could predator and prey, in the evolutionary sense, exist one without the other?

Those *Homo sapiens* who eat meat, but do not hunt, in essence hire livestock producers and processors to do their "hunting" for them. Their meat comes neatly packaged from the butcher with no reminders of what went into that package. Hunters and their families know exactly where meat comes from and what is required to bring it to the table. Dealing death—and blood—are an inevitable part of the equation.

I suspect that only a minority of hunters are introspective enough to delve deeply into themselves and their motivations for hunting. And fewer still are willing to speak or write about what they think, feel, sense, and wonder about on such introspective journeys. But those who dare to embark on such inward analytical journeys know, and share, in the age-old mystique of hunting.

Such journeys, in my experience, are best undertaken while resting, gun in hand, in the sun on a south-facing slope of a mountainside during clear, crisp fall weather. The mood should be abetted by gray jays sweeping in to seek scraps of food left from my lunch, reverberations of a bull elk's bugling on a distant hillside, or perhaps when snuggled deep into a down sleeping bag watching the campfire die into embers and then fade into darkness while dreaming of tomorrow.

To my mind, hunting trophies are not, and should not be, limited to unusually large antlers or horns hanging on the wall of a den or a museum. To me, a trophy is but a tangible reminder of an achievement or, even more, a cherished experience. Only a very few of my trophies hang on a wall—rather they reside in my heart and mind where others cannot see. Where trophies do adorn walls, they should be admired for the animal they represent—and its wild habitats, which are, after all, the hunter's habitat as well. Hunters should return, at least in spirit, to those habitats as they ponder their trophies. Such reverie is the real trophy.

While I worked for the TGD, I killed a half-dozen white-tailed deer bucks, a couple of mule deer, and a pronghorn antelope that were, by any standard of measurement, trophy class. However, because my bride Margaret objected to dead animal parts hanging on the walls in our house (I could not afford the services of a taxidermist), my trophy antlers and horns hung from the rafters of the garage. When I left Texas to go to work for the USFS in West Virginia, I gave the antlers and horns to the Llano County Chamber of Commerce for mounting and display.

I kept only one trophy: a carefully cleaned and bleached piece of skull from a yearling whitetail buck that sported spike antlers as short as my middle finger screwed onto an unvarnished pine board. That genuine trophy was taken by a cherished hunting companion with whom I hunted but once.

The occasion was the Orphans' Doe Hunt in Mason County, Texas, in 1962. The hunt was organized by game warden Gene Ashby, who had a no-nonsense manner with adults coupled with a genuine soft spot for young people. He hoped to encourage more landowners to harvest antlerless deer while introducing some young men without parents—and no opportunity to hunt—to hunting. The meat went to the institutions where those young men lived. It seemed to me a win-win-win situation.

By the luck of the draw, I served as host and guide for a young would-be hunter named Benny. I was waiting for him at the courthouse square in Llano when a chartered bus filled with excited young potential hunters arrived from Austin. Benny was fourteen or fifteen, a short, scrawny kid with close-cropped brown hair and big brown eyes. We drove directly to a secluded spot that provided a safe place for some shooting lessons. Within an hour or so, Benny, using my Marlin lever-action .22 caliber rifle mounted with a Weaver 4-power scope, progressed in his marksmanship to the point where he could, using a rest, consistently knock aluminum cans off a log at twenty-five yards. Then, using my vintage Savage .250-3000 rifle, shooting off a rest, he could knock over three out of five cans at fifty yards. It was pure pleasure watching Benny's excitement as his skills as a marksman improved.

We arrived home just in time for a dinner of chicken and dumplings. Benny didn't say much at dinner, but what he said was always thought-out and politely expressed. After dinner, with just a little help, Benny cleaned the rifles we had fired earlier. My old lever-action Savage was clean and ready for tomorrow's hunt. I was to be Benny's guide.

A couple of hours before daylight, I opened Benny's bedroom door to tell him to rise and shine. He was already wide awake, grinning broadly, lying on top of the covers fully dressed save for his tennis shoes, and cradling the old Savage rifle. The boy was ready—he had gone to bed ready!

By the time the eastern horizon glowed red, Benny and I were in our assigned hunting area seated under a large live oak at the edge of a fallow field. We could just barely make out deer feeding in the open field. I studied the deer through my binoculars and directed Benny's attention to an adult doe about 100 yards away and moving slowly toward us.

As the clock ticked down to legal shooting time a half hour before sunrise, Benny was focused on the deer and had the rifle braced against a mesquite log. I whispered in Benny's ear to wait until she turned broadside and her chest afforded a more certain shot. Just as the deer turned, Benny came down with a classic case of buck ague— in Texican, "buck agar." He trembled and breaths came in gasps. I put my arm around his shoulders and whispered encouragement.

Benny steadied down. Then he simultaneously jerked the trigger while lifting his head in anticipation of the recoil—and missed clean. The doe didn't bolt but came to full alert and stared toward us. Benny repeated his shot—another clean miss. The doe bounded away with her big white tail waving good-bye. Benny began to tremble, bitterly disappointed and seemingly ashamed. He was a tough kid and tough kids don't cry, at least not where anyone else can see. I pretended not to notice.

I talked quietly to him about controlled breathing and squeezing the trigger. I had him aim and dry-fire the rifle a half-dozen times. I assured him that every hunter, including me, missed a shot from time to time. I dug into my hunting pack and pulled out some venison *charqui* and a can of soda pop to recharge Benny's batteries. I told him that eating the dried deer meat would solicit assistance from the Lord of the Hunt. He looked at me quizzically, but he calmed down.

We eased along into the slight breeze from one likely stand to another, spending a half hour or so in each. We saw many deer as the hours passed. Some spotted or smelled us and snorted, threw up their tails, and bounded away. Some did not sense us but were not close enough for Benny's level of marksmanship. A little after midday, we ate lunch and took a short nap, stretched out flat on our backs in the warm sunshine.

Finally, in that magic time between sundown and full dark—the hunter's witching hour—we were working our way back toward my pickup when we came upon a deer standing broadside to us, head down and feeding, less than forty yards ahead. The gentle breeze was in our face and the setting sun at our backs. A mesquite tree provided both cover and a rest for the rifle. Maybe, I thought, the Lord of the Hunt had stepped in. This was Benny's last chance: less than five minutes of shooting time remained.

I saw through my binoculars that it was a yearling buck with spike antlers. Our agreement with the landowner was that Benny could take a doe "on the house." If he killed a buck, I would pay the going price of one hundred dollars, nearly a quarter of my monthly take-home pay at the time.

I looked at Benny, who was braced against the mesquite sighting at the deer without a sign of buck fever. Only a second elapsed between my whispered "shoot" and the crack of the rifle. The deer leaped, spun around, and ducked out of sight into the beeweed and mesquite cover.

Benny began to shake and stutter with frustration. I knew the bullet had struck low in the chest just behind the "elbow"—a perfect heart shot. I signaled Benny to kneel down beside me. I pulled him close and whispered that I thought he had made a perfect shot but we would wait a few minutes before we started looking for his deer. I dug a flashlight out of my hunting pack. We walked to where the deer had been standing. There was glistening red arterial blood—and a lot of it. We followed the blood trail for less than thirty yards, and there lay Benny's deer.

I handed Benny the flashlight and stood motionless as he circled the animal several times. Then he knelt on one knee and stroked the little buck's coat. I crouched next to Benny and put my arm around his shoulders and whispered in his ear, "You can't decide whether you're sad or glad."

He had tears in his eyes, but a smile twitched at the corners of his mouth. "How did you know?"

"That's the way I felt the first time—and every time since. Most hunters don't talk about it. But most hunters—the really good ones,

I think—feel that way. When I don't have that feeling, it will be time for me to quit."

We field-dressed the buck, talking but little in the process. I tied the legs together, slung the deer over my shoulder, and we walked to the nearby ranch road. I told Benny I would tie my handkerchief on a mesquite tree limb to mark the spot, and we would walk a mile or so to the truck and come back for the deer. Benny didn't want to leave his buck—his trophy in the very best sense. So we gathered some dry wood, and I built a small warming fire. I draped my hunting jacket over Benny's shoulders and left.

By the time I got back with the truck, Benny was well past introspection and awe. He jabbered with excitement as we loaded the buck and talked nonstop until we reached the ranch house. I told the rancher what had transpired and I figured I owed him a hundred dollars. I also admitted that I could only pay him twenty dollars at the moment and asked if I could work off what I owed.

The grizzled old rancher adjusted his steel-rimmed bifocals. He looked first at Benny, whose face was all grins—and then at me. A big smile creased his leathery face as he slapped Benny on the shoulder, winked at me, and proclaimed, "Boys, this big ole buck is on the house!"

We visited for a while—hot black coffee for the men and an RC Cola for Benny. He told the old rancher all the details of his hunt; his voice still quivering with excitement. As it was getting late, we said our good-byes. The grizzled but soft-hearted rancher used both of his big callused hands to shake Benny's hand and tousle his hair. As he walked into the house, he pulled out his bandanna and wiped his eyes—dust, no doubt.

Benny and I registered his deer at the TGD's check station and delivered it to the local meat-processing establishment. The packaged meat from the deer killed by the young hunters would be delivered later. I borrowed a bone saw and removed the skull plate with the spike antlers attached and handed the antlers to Benny.

The next morning Benny and I were up before sunrise. After breakfast, we skinned the hide away from the skull that held the spike antlers and scraped and sanded away the remaining bits of blood and flesh. Then we cut a pine board in a triangle and attached the antlers with a couple of wood screws so that Benny could hang his trophy on the wall of his room.

Only then did we eat our breakfast of deer sausage and biscuits and gravy that Margaret had prepared for us. Then, too soon for both of us, it was time to take Benny to the chartered bus. The last time I saw Benny, he was clutching his genuine trophy as he climbed the steps into the bus. He turned back, smiling from ear to ear, and waved good-bye.

I never saw Benny again, though we exchanged letters a few times. Then five years or so after our hunt, a package came in the mail. The package contained Benny's trophy. The enclosed letter told me that Benny (now signing his name "Ben") had finished high school and was off to the Marine Corps. He asked me to keep his trophy safe for him until he could come back and call for it. I wrote back and promised to do just that. I never heard from Benny again. I hope he just forgot about our trophy, but down deep I doubted that.

I still have our trophy—Benny's and mine. I would rather have that trophy than the biggest set of antlers from the biggest buck or the biggest elk I ever killed. I guess there are other ways of keeping score than the B&C's.

BEHIND THE MEAT POLE

In the spring of 2005, a longtime friend and professional colleague, Dr. Michael "Mike" Zagata, executive director of the Ruffed Grouse Society, invited me to address a newly formed sportsmen's group in upstate New York. I drove up from Washington, D.C., where I had been testifying before a congressional committee. Though I felt at home in the countryside (I had done some wildlife research in this area in the late 1960s), I was struck by the dramatic changes in forest conditions that I perceived to have taken place in the four decades since my last visit.

The landscape I remembered had been a mosaic of old fields and pastures intermixed with woodlots or patches of forests with various sizes and species of trees. Hillsides within larger ownerships were covered with hardwood forests on the flats and the south-facing slopes and with mixed conifer and hardwood stands on the north slopes. Stand conditions ranged from fresh clear-cuts to densely stocked second growth to thinned stands to older forests with closed canopies and sparse understory. I remembered small timber mills and log yards along the roads and pulp mills spewing smoke and somewhat foul smells off in the distance.

Forty years earlier, the area had been a Mecca for hunters in pursuit of white-tailed deer, ruffed grouse, and wild turkeys. Deer numbers, in particular, had exploded in the first half of the twentieth century as marginal agricultural operations were abandoned and the mixed hardwood/conifer forest began to reclaim its ecologically ordained place in the landscape. The situation for game managers had gradually shifted from protecting and reintroducing white-tailed deer to dealing with overpopulations causing damage to farms and young forests. Late-winter mortality among deer was increasingly common, especially during long winters with heavy snow accumulations. These landscapes were the scenes of some of the first of the "doe wars" in the 1940s and 1950s, when wildlife biologists tried to convince recalcitrant deer hunters that the era of deer scarcity had given way to an era of abundance and then overabundance with resultant damage to habitat.

Now, nearly a half century later, the landscape had changed dramatically. The mosaic of varied habitat conditions that had produced such high deer populations was dramatically altered. Mostly young and mid-succession forests now stretched from horizon to horizon with many fewer breaks in the canopy. The essentially closed canopies thwarted sunlight from reaching the forest floor, thereby reducing the primary source of food for deer. Deer had, in many cases, browsed away what vegetation that did exist as high as they could reach standing on their hind legs. Known as a "browse line," it is a sign of a deer herd depleting its habitat.

The primary question at the sportsmen's meeting was "Where have all the deer gone?" I gave my pitch. Habitat conditions had changed dramatically over the past half century and could now support but a fraction of what existed "back in the day." Local biologists had, of course, routinely made the same points. But in the wildlife management business, professional prophets rarely have honor in their own countries.

As I spoke, a few people in the audience nodded knowingly in agreement. A few seemed puzzled. Most were polite enough, but their body language of set jaws, arms crossed across their chests, slumping down in their chairs, and the gentle shaking of their heads from side to side spoke volumes. I was reminded of the ten years I spent with the TGD trying to convince landowners and hunters to control deer numbers to assure that habitats were not overused. Some things never change.

One old-timer—even older than I—begged to differ with the points I had tried to make. In his mind, the entire cause of declines in deer numbers was that hunters were killing too many young bucks and way too many does (antlerless deer). He believed such hunting pressure had caused a steady deterioration in the number of bucks and a decline in antler size. He had a slideshow to back him up—after all, pictures can be worth a thousand words.

His slideshow was a series of pictures of the "meat pole" at his hunting camp, showing the bucks killed each year over a half century. The antler sizes of bucks hanging from that meat pole had obviously declined over the years. Just before he sat down, he looked at me and asked "what I thought of them apples."

Sometimes, Providence affords a teachable moment, and this was one. I asked him to quickly run through the slides again. I asked the audience to ignore the deer and the hunters and the areas immediately in the background and rather to concentrate on the hillsides. The result was akin to an old-time silent movie. They could see forest stands growing from bare ground to shrubs and seedlings to young trees to a pole stand, under which there was little sign of ground vegetation or leaves and twigs as high as deer could reach. Frame by frame, the mosaic of old fields and woodlots in various stages of cutting and recovery morphed into wall-to-wall stands of early and mid-succession forests with little understory.

Cause and effect was plain to see. As the number and size of bucks hanging from the meat pole declined—perhaps also related to a too-heavy harvest of antlered deer—the forests and scattered openings in the background changed from highly productive habitat for deer, grouse, and turkeys to its current, much less productive state. Most of the audience got the picture—no pun intended. I thought my protagonist did likewise, but he wasn't about to admit it. I asked him for permission to copy the slides—at my expense. The answer was no.

Some animals, deer among them, thrive on a mixture of forest conditions. They need openings (including early forest succession) and the "edges" between cover and openings for forage as well as more mature forests for cover from predators (including hunters) and protection from extreme weather. Openings in the forest canopy, which are commonly transitory in unmanaged landscapes, are naturally created through disturbance—wildfires, blowdown, insect and disease outbreaks, and flooding, among others. Or the disturbance can result from human actions such as clearing for agriculture, logging, application of herbicides, controlled burning, and intentional creation and maintenance of forage areas for wildlife. After being disturbed, lands suitable to forests conditions "want to be forests," and most will, once again, tend toward becoming forests if allowed to do so.

Natural disturbances that occur over long timeframes are more or less random, varying in kind, timing, effect, extent, and location. On the other hand, human-induced disturbances are, to large extent, controlled for a variety of reasons, including manipulation of wildlife habitat.

In this region, abandonment of active agriculture and forestry, for whatever reasons, commonly leads, over time, to mid-succession forests with little understory. Animals that thrive in a mosaic of openings and forest stands of various ages decline in numbers as forests reach sapling-pole stages. If resident deer numbers are not controlled to be in balance with changing habitats, deterioration in physical condition and reproductive rates often ensues.

The trick to understanding the connections between habitat conditions and wildlife numbers—and discerning appropriate approaches to the management of both animals and habitats—comes from, in this example, "looking behind the meat pole." Those relationships have been clear to wildlife biologists since the 1930s. Yet wildlife managers over eighty years later still have trouble persuading many hunters—and politicians—to adjust their hunting goals to keep populations in balance with changing habitat conditions. It seems that some things never change.

BELIEVE IN MAGIC

In 1994, I was honored to be inducted as a Professional Member into the Boone and Crockett Club, among the oldest organizations in North America dedicated to furthering the conservation of wildlife. After accomplishing great things for conservation in North America between 1887 and 1916, B&C gradually dropped out of the forefront of national conservation efforts and focused on keeping records of North American big game trophies. Then William "Bill" Spencer was elected the president in 1983 and rallied the venerable organization to vigorously reengage in conservation affairs.

Daniel "Dan" Pedrotti, oil man, rancher, and dedicated conservationist from South Texas, followed Spencer as the Club's president some fifteen years later. Pedrotti was, by any measure, an accomplished entrepreneur and a truly big man in terms of achieving much in private enterprise and in the business of conservation. He graduated from Texas A&M and had served as a U.S. Air Force flying officer.

Dan took over the presidency with a vision of what this old, prestigious, and now revitalized organization might accomplish next. He had watched and analyzed the emergence of successful private-sector conservation organizations and their various *modus operandi*. He noted that old-line conservation groups such as the Wildlife Management Institute, Izaak Walton League, National Wildlife Federation, National Audubon Society, International Association of Fish and Wildlife Agencies, and others had been, in recent years, joined by a number of conservation organizations that focused upon a previously ignored niche: sportsmen's interests in individual hunted species. Among these newly flourishing organizations were the Rocky Mountain Elk Foundation, Wild Turkey Federation, Quail Unlimited, Mule Deer Foundation, Ruffed Grouse Society, Foundation for North American Wild Sheep, and Whitetails Unlimited, among others. He noted the significant overlaps in the memberships of these organizations and with professional organizations that were involved in the work of natural resources conservation such as The Wildlife Society, Society for Range Management, American Fisheries Society, Society of American Foresters, and others.

While the total membership of such organizations was, in total, huge, their overall influence was much less than Pedrotti believed it could and should be. If these organizations could be gathered under one umbrella to take on the significant conservation issues, their goals would be more likely to be achieved.

The Regular Members of B&C included some of the most influential men and women in America. In addition, its Professional Members included many of the most experienced and accomplished professionals in the natural resources arena—folks from state and federal conservation agencies, academia, and the world of politics who could be called upon for guidance and help in bringing his vision to fruition.

Pedrotti saw the potential of the "bully pulpit" afforded by the Club's presidency in making his dream a reality. He developed a strategy—with assistance from Steve Mealey, B&C's executive director, who had just retired after a distinguished career as a wildlife biologist with the USFS, and from various Professional Members—to convene a meeting of the leaders of the varied organizations at the North American Wildlife and Natural Resources Conference, held in Washington, D.C.

The invited representatives sat in two rows around the longest conference table I had ever seen. Most were there in response to Pedrotti's invitation. Others were there at the behest of friends and colleagues who were Professional Members of B&C. Some came with an open mind. But just as many dedicated warriors of North America's conservation movement seemed jaded with cynicism, some exuding pessimism.

Pedrotti made his pitch. There was no immediate enthusiasm or support for Pedrotti's dream. Some attacked the proposal. Some argued that the time was not right. Others said they already had a full plate. Some enumerated why any such effort would inevitably fail—after all, it had been tried before. Clearly, many in the room could not see the potential of such an effort or were more interested in protecting their own turf and the welfare of their own organizations.

At the time, I held no office in any of the organizations represented around the table, though I was a member of several and had previously held office in more than a few. I had been invited to the meeting by Pedrotti as an observer. Pedrotti's vision—which seemed to me logical, doable, and desperately needed—was dying the proverbial death from a thousand cuts.

Then one of the younger chief executive officers present, dressed impeccably in a three-piece suit, stood up and made a remark that I considered disrespectful to my friends and colleagues, Pedrotti and Mealey. I had not been invited to speak and had not intended to do so. But his remarks brought me to my feet. I can't remember my exact words. But what follows approximates what became known in B&C circles as Jack's "I believe in magic" speech. It went something like this.

"Those seated around this table can, with justification, think of yourselves as the leaders of the wildlife conservation movement in North America. I have heard many reasons why a more united, better coordinated, mutual effort to guide more effective conservation efforts in North America won't or can't work. If those who preceded us in what Theodore Roosevelt called 'the arena' had behaved in such fashion, there is a good chance we wouldn't be gathered here today. Most of the organizations represented here today wouldn't even exist. And, just as likely, there wouldn't be much wildlife or many wild places to worry about. Those pioneers believed that magic could and would happen when the right people rallied around a collective vision with determination to make that vision a reality. Those fledgling conservationists believed in their cause—and, more importantly, in themselves. And they believed in the magic that the cause of wildlife conservation could work in the body politic. They believed in the people's ability and willingness, if and when properly led, to meet a worthwhile challenge. I, for one, believe in magic—the magic that can emerge from the right people in the right place at the right time with a worthwhile cause. Don't dither. Don't fuss. Don't nitpick. Don't give up before you start. For God's sake, trust the magic!"

A little embarrassed, I sat down. It was very quiet. Then the reticence began to crumble as, one by one, individuals stood up and spoke up, some quite eloquently, in support of Pedrotti's dream. By the time the meeting was over, it was agreed that B&C would organize and host a meeting of the heads of the organizations represented in the room and appropriate others who saw the potential in Pedrotti's dream; the several-day workshop would be held at the Club's headquarters in Missoula, Montana. Suddenly, even those who had initially been so negative spoke up to give Pedrotti's proposal their support. It was as if a dam had broken.

So magic happened—or, more accurately, magic started to happen.

Several months later, the meeting took place in Missoula. The first day much of the trepidation and reservations previously expressed in Washington resurfaced. At the end of that day, Pedrotti and Mealey convened a small cadre of B&C members who were present to face up to the reality that the effort was, once again, on the verge of jumping the tracks. The "old magic" wasn't working the way we had hoped, and several participants were talking of leaving.

We concluded that we needed an expert facilitator with no investment in the outcome and no ax to grind. My bride, Kathleen "Kathy" Hurley Thomas, had served in the federal government's Senior Executive Service in Washington, including the Department of the Navy, the USDA, and the USFS, from which she had recently retired as deputy chief for administration. Being an expert facilitator was part of her skill set, and she was among the very best I had ever seen ply their skills over my fifty-year career. Since Kathy was my wife—and it was assumed that I had some influence—I was assigned to ask if she would take on the job of facilitation, despite the short notice and a situation trending strongly toward failure. It really didn't take much persuasion. She was already a believer, and she liked and admired Dan Pedrotti and wanted very much for him to succeed.

She divided the attendees into smaller working groups to address the "sticky" questions already formulated by the larger group. She quickly recruited B&C's attending Professional Members to serve as facilitators for each working group, including: Dr. Dan Pletscher of

Conservation leaders at the first meeting of the American Wildlife Conservation Partners, held the Boone and Crockett Club's headquarters on August 8, 2000.

the University of Montana; Dr. Kaush Arha of the Wyoming Department of Fish and Game; Butch Marita, a retired USFS regional forester; Dr. Greg Schildwachter of Western Timber Association; and Dr. Lisa Flowers, the Club's director of education. The Rocky Mountain Elk Foundation provided two experienced facilitators—Tracy Scott and Dave Torrell.

I had the job of moving between groups to listen, analyze, and report back to Kathy about progress and emerging sticking points. After she processed and correlated the information from all the working groups, she would send me back to announce a short break. The facilitators gathered and coordinated their "making of magic" in order to move the group discussions forward—and salve the egos involved—so that all the groups would arrive at somewhere close to the same place at roughly the same time.

The process reminded me of what my high school basketball coach called "rat ball"—or, in techno-speak, "rapidly evolving ad hoc organization." Kathy and the facilitators demonstrated agility in doing "whatever works" while "believing in magic." To the amazement of many of those present, barriers crumbled and resistance faded. When the entire group reassembled and the working groups reported, the almost unanimous decision was made to form an umbrella group in wildlife conservation.

I thought of the wisdom of the old Sioux chief in the movie *Little Big Man*: "Sometimes the old magic works and sometimes it doesn't." This time the old magic worked.

Dan Pedrotti was so impressed with Kathy's performance that he nominated her to be a Professional Member in B&C. That was a real honor for a woman who was more at home in the big city—she called herself "subway girl"—than in the outback and who had never fired a gun at an animal in her life—and had no intention of ever doing so.

Pedrotti's concept had become reality. Dr. Rollin "Rollie" Sparrow of the Wildlife Management Institute; R. Max Peterson, chief emeritus of the USFS and director of the Association of Game and Fish Commissioners; and B&C representatives Dan Pedrotti, Robert "Bob" Model, and Steve Mealey would organize the first formal gathering of the new American Wildlife Conservation Partners, to take place at the next North American Wildlife and Natural Resources Conference. The die was cast. Against the odds and despite a rocky start, the wildlife conservation coalition was a reality. Her job done, Kathy joined me in the seats set aside for observers. Now we had no role in what was going on around the table, but we had been there in the beginning and had been of some help. We were proud of that.

What went on confirmed my belief that conservationists all too often snatch defeat from the jaws of victory by thinking small, focusing on preserving power and influence in organizations and individuals, and refusing to believe in the power of magic. "Magic" can happen when the right people come together at the right time with the right mindset. "Come, let us reason together" is ancient wisdom that has withstood the test of centuries. The organizers of the new American Wildlife Conservation Partners did exactly that, and it paid off.

Good on Dan Pedrotti!

JUST A PILE OF ROCKS

My father-in-law, John Hurley, once made an offhand comment that set me to thinking about mountains. A sensitive, perceptive, intelligent, and well-read man, John was at heart a Bostonian Irishman (an altar boy in his youth) with an undiluted accent to testify to his heritage. As a wildlife biologist might put it, his preferred habitat was Boston and Washington, D.C., with time out for service in the Pacific in World War II.

Just shy of his ninetieth birthday, he and wife Marie moved from Maryland to the "wilds" of Montana to settle across the street in Florence from their oldest daughter, my wife Kathy. Somewhere in the process, he grumbled that "Montana is almost in Canada." Beyond that, Florence is a one-stoplight town in the heart of the Bitterroot Valley south of Missoula. The Bitterroot Valley is delineated by mountain ranges: the Sapphires to the east and the Bitterroots to the west. The mountains are awesome at any time of year, but most of all in late October when the first winter snows form swaths on the upper slopes.

In late October, Kathy, her parents, and I attended a monthly neighborhood luncheon at a restaurant in Stevensville, a two-stoplight town a few miles up the Bitterroot Valley. After lunch, as I waited by the car for my passengers, who were still visiting with neighbors they had not seen for spell, I gazed up at the snow-covered peaks of the Bitterroot Range shimmering in the midday sun.

When John walked up. I remarked upon the grandeur of the upper slopes newly clad in snow. John grunted, "I don't think they're beautiful. They're foreboding—just a big pile of rocks!"

I was taken aback. I concluded that the difference in how we viewed the mountains was rooted in our different backgrounds and lifetime experiences—and in our "knowing." I had known mountains intimately—around the globe, up close and personal—over many decades and viewed them as awesome, beautiful, inspiring, and home to unique forms of wildlife. He had no such experiences and perceived the same mountains as sinister and threatening.

The beautiful beginning of winter in the high lonesome in Montana coincides with the hunting seasons for mule deer and elk, which remain up high until the deepening snow forces them down into the valley, first around the edges, where they sometimes mix with white-tailed deer that remain in the valley floor environs year around. Only a few old bulls and big bucks hold out in the high country for just a while longer.

Some hardy humans trek those mountains during the summer months. Some hike, carrying their gear on their backs. Others are on horseback, leading packhorses or mules that carry their *accoutrements*. More and more travel via gasoline-powered trail bikes or four-wheelers. The travelers have their individual objectives in terms of destinations and experiences sought. The mountains afford increasingly rare opportunities to experience and learn from the silence and solitude so closely coupled with the mountains' inherent magnificent beauty. After all, there were mountains before there were cathedrals, and there will be mountains when all man's creations have crumbled.

Up high on the mountainside, on a cloudy night, one can marvel at unadulterated darkness. On cloudless nights, stars blanket the heavens so thickly and shine so brightly that it seems almost possible to reach out and gather them in by the handful. The mountains sing always, even in the darkness: the wind in the trees, water running, coyotes and just maybe a wolf or coyote howling, bats flitting, night birds calling and flying about. If you lie quiet and listen—really listen—to the night sounds, you can hear your heart beating, the blood coursing in your veins, and your breathing. The smell of wood smoke from the dying fire comes and goes with shifts in the breeze.

Some nights bring cold winds and slow, steady rain. The near-perfect sensation is squirming deep into a down sleeping bag and listening to raindrops drumming lightly on a tightly stretched tarp or tent—ointment to salve the sorest soul. Sometimes a storm brings excitement in the form of lightning and wind and rain. Sometimes the wind is so strong that pilgrims wonder mightily if the tent pegs will hold. Some still mornings reveal a blanket of snow that fell quietly during the night.

In October, the mountains belong to those hunters hardy enough to penetrate deep and climb high, willingly accepting the risks that can come along with rapidly changing weather and the chance of deep snow or perhaps serious mishaps. Such late-fall visitors march to a different drummer than those who come in the summer.

Some hunters leave the USFS's maintained roads and follow game trails or forge trails of their own in pursuit of mule deer or elk— or, maybe just as important, solitude. The hunters I have known well, admired most, and felt the closest kinship sought much more than prey. Their feelings evolved over decades and required the too-rare combination of an open heart and open mind.

I came to believe that such introspective journeys of the mind are best undertaken high in the mountains. When Bill Brown, my mentor, friend, and companion on many mountain wilderness trips in Oregon, Washington, Idaho, and Montana, referred to the mountains as "the high lonesome," he spoke not only of altitude but a state of mind as well.

How could I explain all that to John Hurley, who was born and raised, prosperous and happy, in the cities of Boston and Washington? Why were the mountains more to me—so much more—than a foreboding pile of rocks? They were a gateway to wonderment and a means of expanding mind and soul. I knew I could not explain that to John Hurly so that he would understand. Why? Simply because I had been there so many times while he had never been there, not even once. I knew the high lonesome, up close and personal. He did not and never would.

Long ago I encountered a passage by René Daumal that came close to expressing how I felt about the high lonesome. I copied the poem on a sticky note and stuck it to the edge of my computer monitor. I've had to replenish the note several times over the years.

> You cannot stay at a summit forever; You have to come down again. So why bother in the first place? Well, just this: What is above knows what is below, but what is below does not know what is above. One climbs, one sees, one descends, one sees no longer, but one has seen. There is an art of conducting oneself in the lower regions by the memory of what one saw higher up. When one can no longer see, one can at least still know.

I know, but admit to no one, that I will never again go to the high lonesome. The spirit is willing, even deeply longing, but the accumulating years and old injuries and increasing health problems make such a visit improbable. I make no complaint. I live now in a beautiful valley. The ground is flat, the roads are paved and plentiful, the meat market is nearby, and the air is rich in oxygen. Even as I miss the high lonesome, I retain that great treasure in my heart and mind. For, after all, I know what's up there. And I go there often, if not now in person, in reverie.

I remember the trails, high mountain lakes, cold clear air, cold rushing streams, well-trained horses, huge yellow-bellied ponderosa pines, towering Douglas firs, waving stands of bunch grasses, stands of willows along mountain streams, trout rising, profound silence, elk bugling, and the totality of it all—the sheer beauty and wonderment.

I remember places wild and free and I know that, for a little while (it seems now such a very little while) I was part of them. I made a life's work of trying to understand wildlife and wild places in a wild world too fast disappearing. I did what I could to understand and protect that world, at least for a while longer, so that others—including those yet to be—could sense, feel, experience, and know some of the marvelous things I have been blessed and privileged to know. All of that lies up there, high in the mountains in those "foreboding piles of rocks," waiting patiently for those who choose to seek.

In the early 1970s, when I organized and directed the USFS's research effort dealing with urban forestry and wildlife, we established study areas in the cemeteries, parks, river bottoms, and neighborhoods of Boston—areas that John Hurley would know, appreciate, and find comforting. Since I remained largely ignorant of the glories of Boston, I might dismiss his beloved habitat as a bunch of old buildings, a jillion people, and a very foreign land where the natives had a very strange manner of speech. Why? I lacked "the knowing" of the place, and that precluded deeply held affection and appreciation. So, at bottom, I was no different from John Hurley. We were simply familiar with, at home in, and inspired by different habitats.

After all, in the end, it is all in the knowing.

EPILOGUE

In early January of 2015, when I sat down in my den in my home in Florence, Montana, to write this closing chapter, it was bitterly cold and snowing—days were now short and nights long. Winds were howling down from the Bitterroot Mountains, swirling the falling snow. I tuned my old radio to a station that played golden oldies, as is my custom when puttering in my den late into the night—which is ever more common now as bones ache and sleep evades.

"September Song" was playing. The song's refrain, "the days dwindle down to a precious few," reflected my mood. I sensed how late it is in the autumn of my life. My days are indeed dwindling down to a precious few.

The last three-quarters of a century has been one long adventure for me, even considering the rough patches. If it were possible, I would—without hesitation—sign on for another tour. I don't know if heaven exists, or what it might be like if it does. But for me, one sort of heaven would be simply doing it all over again.

My profession as a wildlife biologist and conservationist made it possible for me—and my colleagues, past, present, and future—to strive to responsibly tend wild places and wildlife under evolving circumstances. In making that choice, I left "preservation" to others—not because I considered preservation unworthy but rather because I recognized such efforts could, at best, save only small segments, however magnificent, of the earth from exploitation and then only briefly. I reasoned that *Homo sapiens*, like all species, will exploit its environment in order to live, survive, and prosper.

I believed that how that exploitation took place was a key to assuring that wildlife, and relatively wild places, retained a place in an ever-changing world. Striving for knowledge and then weaving that knowledge into packages to guide informed management of natural resources has been my passion.

Once, when I was winding up my tenure as chief of the USFS, a reporter asked me what I would want my epitaph to say. I thought about this for a while and then told him: "He was a good husband, father, and grandfather. He was a professional conservationist of some note. He strived to be honest and did his best. He freely shared what he knew and thought. And he was a fine hunter."

I could tell that the reporter was with me until the final sentence. He looked up from his notepad, puzzlement on his face. Then he smiled, thinking that I was joking. I wasn't. It was a teachable moment, and old professors never pass up teachable moments.

Actually, my epitaph will read simply, "Jack Ward Thomas. Born: September 7, 1934. Died: *Whenever*." You come into this world with nothing and go out the same way. It's the in-between that counts, and that should speak for itself.

It is late now—late at night and late in life. I am deeply engaged in a battle of old age exacerbated by pancreatic cancer (a particularly nasty brand of cancer)—likely the final fork in my trail. But life is still good and, strangely enough, I am more curious than afraid.

The fire in the fireplace has died down to embers. The room suddenly seems chilly. It is time for me to stop. As I lie back in my easy chair, waiting for sleep to come, it will be time to ponder the next fork in the trail—what, where, and when.

Surely, there is still an adventure or two somewhere along the trail between here and that final fork. But now it is time to sleep, "perchance to dream." My dreams, I hope, will be of the high lonesome and dear ones and comrades who have gone before—and of past and, just maybe, future forks in the trail.

PUBLISHER'S NOTE

The Boone and Crockett Club would like to recognize several individuals whose hard work, diligence, and support—both financially and professionally—made this book project possible. As with most publishing projects, this was indeed a group effort.

First and foremost we must thank the author, Jack Ward Thomas. He is a true conservationist, biologist, and hunter beyond reproach. His contributions to our natural world will be felt for generations to come. Jack's choice of the Boone and Crockett Club as the recipient of his journals and manuscripts is an honor for which our publishing program is deeply indebted. This generosity was facilitated by two Boone and Crockett Club Honorary Life Members, John P. Poston, and Daniel H. Pletscher. These two gentlemen initially met with Jack and through several conversations approached the Club's publications program with the idea that we should publish this trilogy of Jack's journals and memoirs. Who better to publish this great conservationist's work than Boone and Crockett—founded in 1887 by Theodore Roosevelt and George Bird Grinnell, with over 120 years of publishing experience. We are indebted to John and Dan for their foresight and their tenacity.

A special acknowledgment and debt of gratitude needs to be extended to John Poston for his financial support that made this project happen. It would not have been possible otherwise to move this project from an idea in the fall of 2014 to three finished books by the summer of 2015. John's support, reflecting his friendship and respect for Jack Ward Thomas, allowed us to get the ball rolling on a much faster schedule than anyone could have hoped for. We can only hope to have a faithful friend like John Poston in our lifetime!

The Boone and Crockett Club has a long history with publishing books about outdoor adventure, conservation, and hunting. The Club's publishing program is overseen by myself and Jeffrey A. Watkins (Cartersville, Georgia). Jeff's enthusiasm, suggestions, and oversight were instrumental in developing the publishing concept. Most importantly we are fortunate to have the assistance of Julie Tripp, B&C's director of publication. Julie is the foundation helping Jeff and me develop a working plan to get the project underway and see it through to completion. Julie especially enjoyed her trips down the Bitterroot Valley from B&C headquarters in Missoula to visit with Jack and his wife Kathy throughout the publishing process. Time was spent pulling photographs, going through edits, selecting authors to write the forewords, but more importantly, just "shooting the breeze" with Jack. If only we had had a stand-around fire to set the stage for those conversations!

There are a few others whose hard work made these books happen, and we'd be remiss not to mention them here. Fellow Texican Alison Tartt did an exemplary job editing and arranging the three manuscripts into their final form. B&C Lifetime Associate Hanspeter Giger (Charlotte, North Carolina) volunteered countless hours of his time providing a final read-through of the books before they went to the printer.

Thank you to everyone who helped us in the creation of this worthwhile project. It is our hope that these books will entertain, inform, teach, and induce reflection in the readers about the natural world we all have the privilege to enjoy.

Howard P. Monsour Jr., M.D.
CHAIRMAN, B&C PUBLICATIONS COMMITTEE
HOUSTON, TEXAS

AUTHOR'S ACKNOWLEDGMENTS

This book project has been decades in the making. Along the way, I was blessed with two strong women in my life who need to be acknowledged for their part in this great process.

Farrar Margaret ("Meg") Thomas and I were married in June of 1957, the week after I graduated from Texas A&M. We were full partners for the next thirty-six years—including, in the beginning, some very lean years when she taught piano lessons and I worked on weekends and holidays at part-time jobs to make ends meet. She never, not even once, complained. She also taught music in the public schools, and privately, to enhance our meager income.

Though she was no hunter, she seemed to relish going hunting with me. She seemed, especially, to enjoy my hunting bob-white and blue quail. I think she was partial to the bird dogs. When, in 1977, I asked her what she thought about leaving Texas so I could take a job with the U.S. Forest Service in West Virginia, she said she would go anywhere if she didn't have to cook and eat venison "nearly every damn day." We were off to West Virginia—and she still cooked and ate a lot of venison. But for the first time in our married life she didn't have to. That made all the difference.

She died of cancer in 1993 in Washington, D.C. Though severely ill, she insisted that "we" accept President Clinton's offer for me to be the thirteenth chief of the U.S. Forest Service. The president referred to Meg as the "First Lady of American Conservation."

In 1996, I was married to Kathleen Hurley Connelly, who worked with me in the Washington office of the U.S. Forest Service as the Deputy Chief for Administration. She was not a "country girl" and even referred to herself as "Subway Girl." Though no hunter herself, she relished accompanying me on hunting trips in the United States and foreign lands. Shortly after we married, we moved to Missoula, Montana, when I accepted an offer to become the Boone and Crockett Professor of Wildlife Conservation in the College of Forestry at the University of Montana. I enjoyed teaching and dealing with graduate students. Kathy served us all as "Mom in Residence." And thoroughly relished the role.

I had a number of bosses over my forty-year professional career. Those bosses—every single one—guided me with a loose rein, which was bound to be, from time to time, a bit nerve-racking. They, in general, gave me the opportunity to "be all that I could be" in both the professional and personal sense. Over a half-century career, my supervisors—in both wildlife research and land management—guided me well and supported me, while "looking the other way" on numerous occasions. There are too many such colleagues to mention by name, but we both know who you are.

My idea of heaven would be to, simply, do it all over again.